Replanting the Tree of Life

**Towards an International Agenda for
Coconut Palm Research**

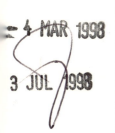

the
UNIVERSITY
of
GREENWICH

Replanting The Tree of Life

Towards an International Agenda for Coconut Palm Research

by

Gabrielle J. Persley

C·A·B International

In association with

ACIAR
Australian Centre for International Agricultural Research
and

Technical Advisory Committee of the Consultative Group on International
Agricultural Research

C·A·B International
Wallingford
Oxon OX10 8DE
UK

Tel: Wallingford (0491) 32111
Telex: 847964 (COMAGG G)
Telecom Gold/Dialcom: 84: CAU001
Fax: (0491) 33508

In association with:

ACIAR
(Australian Centre for International Agricultural Research)
PO Box 1571
Canberra
Australia 2601

Tel: 61 6 2488588
Fax: 61 6 2573051
CGNET/Dialcom: 57: CGI034

and

Technical Advisory Committee of the Consultative Group on International
  Agricultural Research
TAC Secretariat FAO
Rome
Italy

Tel: 39 6 57973306
Fax: 39 6 57973152
CGNET/Dialcom: 57: CGI501

A catalogue record for this book is available from the British Library

ISBN 0 85198 815 6

Typeset by Alden Multimedia Ltd,
Printed and bound in the UK by Redwood Press Ltd, Melksham

Contents

Foreword

This document represents the outcome of an investigation into the need for additional international support for coconut research, undertaken by the Australian Centre for International Agricultural Research (ACIAR), on behalf of the Technical Advisory Committee (TAC) of the Consultative Group on International Agricultural Research (CGIAR). The CGIAR is an informal association of countries, bilateral and multilateral development agencies and private foundations. Its cosponsors are the World Bank, the Food and Agriculture Organization, and the United Nations Development Programme.

During 1988–91, ACIAR undertook consultations with interested parties in coconut-producing countries, development agencies and research institutions worldwide. As part of these consultations, the possibility of an international initiative on coconut research was discussed at the Asian and Pacific Coconut Community (APCC) annual technical meeting in Bangkok in May 1989. ACIAR also convened a small working group of coconut researchers in Singapore on 13–14 September 1989, with the cooperation of the Asian and Pacific Coconut Community and the International Development Research Centre.

TAC considered three papers prepared by ACIAR in 1988, 1989 and 1990. These summarized the position of coconut in the context of the world fats and oils market; the importance of coconut as a subsistence and a cash crop in producing countries; current research efforts; future research needs; which of these needs were appropriate for international support, and their priority; the ideal constitution of an international research initiative on coconut; and various institutional options by which an initiative could be established. It was decided early in the process that the information presented to TAC would be published in order to make it available to a wider audience with interests in research on perennial tree crops. The present document synthesizes the three documents prepared by ACIAR and TAC's own deliberations on coconut research. It also makes recommendations as to the next steps required to foster additional support for coconut research.

Since TAC first commenced its deliberations on coconut, the CGIAR has expanded its coverage to include forestry and agroforestry. The International Council for Research on Agroforestry (ICRAF), and the newly established Centre for International Research on Forestry (CIFOR) are new members of the CGIAR network of international agricultural research centres. It is appropriate that TAC and the CGIAR should consider how to address the research needs of coconut, as a specific case of a multipurpose tree widely grown by small-scale farmers throughout the tropical regions of the world. TAC will remain involved in discussions on coconut research in terms of the expanded scope of the CGIAR system, and the institutional mechanisms most appropriate for international support for coconut research. TAC welcomes the initiative of the International Board for Plant Genetic Resources in the establishment of a worldwide coconut genetic resources network, as a first step towards establishing a broader international initiative on coconut research.

A. McCalla
Chairman
CGIAR Technical Advisory Committee
University of California
Davis, CA USA

G.H.L. Rothschild
Director, ACIAR
Canberra, Australia

Acknowledgements

This document synthesizes information obtained from many diverse sources, and I am grateful to all who gave of their time and experience to the consultations initiated by ACIAR and TAC on coconut research. I am especially grateful for several helpful discussions, on the activities of national research institutes and future research needs, held with S.N. Darwis (Indonesia), R. Mahindapala (Sri Lanka), M Nair (India), G. Santos (Philippines), and P. Turner (Papua New Guinea). P.G. Punchihewa and S. de Silva of the Asian and Pacific Coconut Community also provided much additional information on the activities of APCC and its member countries. The programmes of various bilateral and international research and development agencies were described by A. Bennett, L. Brader, M. de Nuce de Lamothe, O. Dufour, J. Engels, M. Foale, D. Meadows, N. Mateo, J. Palmer, P. Perret and R. Smith. Their contributions are much appreciated.

Special thanks are due to the following persons for their contributions in their specific areas of expertise: D. Adair and D. Etherington on postharvest aspects; E. Cocking on tissue culture; M. de Nuce de Lamothe on coconut breeding; R. Duncan on future trade prospects; H. Harries on coconut genetic resources; M. Foale on coconut-based systems; D. Hanold and J. Randles on coconut diseases; L. Jones on biotechnology; D. Meadows on coconut agronomy and breeding; C. Wing on information systems; and M. Way on insect pests.

The various reviews of oil crops research prepared by A. Green for the European Commission, the Institut de Recherches des Huiles et Oligineaux (IRHO), and the World Bank in the 1980s, and the documents prepared by P. Turner for the FAO Inter-Governmental Group on Oilseeds in 1991 contain much valuable information, and these have been drawn upon here. The oilseed data provided by C. Mielke of Oilworld is also gratefully acknowledged.

The manuscript was discussed by participants at the international workshop on coconut genetic resources, held in Cipanas, Indonesia, 8–11 October, 1991. I am grateful to the participants and to M.A. Foale, F.W. Howard, J.G. Ohler, R. Smith, P. Turner and M. Way for their subsequent detailed review of the manuscript.

Financial support for the study was provided by TAC and ACIAR. The advice provided by A. McCalla, M. Arnold and J. Monyo from TAC is gratefully acknowledged. I would also like to express my thanks to the following for their support and assistance: G. Rothschild (Director) and P. Ferrar of the Australian Centre for International Agricultural Research, P.G. Punchihewa and S. da Silva of the Asian and Pacific Coconut Community, N. Mateo and C. Wing of the International Development Research Centre, P. Greening, H. Marinaccio and T. Williams of the International Fund for Agricultural Research, and C. Bonte-Freidheim and P. Ballantyne of the International Service for National Agricultural Research. Special thanks are due to Arlene Slijk, and Pam Van Den Heuvel for their skilful technical assistance in preparing this document.

G.J. Persley, 1992

Acronyms and Abbreviations

AARD	Agency for Agricultural Research and Development (Indonesia)
ACIAR	Australian Centre for International Agricultural Research
ADB	Asian Development Bank
AICCAIP	All-India Coordinated Coconut and Arecanut Improvement Project
APCC	Asian and Pacific Coconut Community
BUROTROP	Bureau for the Development of Research in Tropical Perennial Oil Crops
CGIAR	Consultative Group on International Agricultural Research
CIDA	Canadian International Development Agency
CIFOR	Centre for International Forestry Research
CIS	Commonwealth of Independent States
CRI	Coconut Research Institute (Sri Lanka)
CSIRO	Commonwealth Scientific and Industrial Research Organization (Australia)
CTA	Technical Centre for Agricultural and Rural Cooperation (ACP-EEC Lomé Convention)
EC	European Community
EMBRAPA	Empresa Brasileira de Pesquisa Agropecuaria (Brazil)
ESCAP	Economic and Social Commission for Asia and the Pacific
FAO	Food and Agriculture Organization of the United Nations

GTZ	Deutsche Gesselschaft für Technische Zusammenarbeit (Germany)
IBPGR	International Board for Plant Genetic Resources
ICA	Instituto Colombiano Agropecaurio (Colombia)
ICAR	Indian Council of Agricultural Research
ICRAF	International Council for Research in Agroforestry
IDRC	International Development Research Centre (Canada)
IFAD	International Fund for Agricultural Development
IFAR	International Fund for Agricultural Research
INIBAP	International Network for the Improvement of Banana and Plantain
IRETA	Institute for Research, Extension and Training in Agriculture of the University of the South Pacific
IRHO	Institut de Recherches pour les Huiles et Oleagineux (France)
IRRI	International Rice Research Institute
MARDI	Malaysian Agricultural Research and Development Institute
NCDP	National Coconut Development Programme (Tanzania)
NRI	Natural Resources Institute (United Kingdom)
ODA	Overseas Development Administration (United Kingdom)
PCA	Philippines Coconut Authority
PCARRD	Philippines Council for Agricultural Research and Natural Resources Development
PNG-DAL	Papua New Guinea Department of Agriculture and Livestock
PPK	Pusat Penelitian Kelapa (Indonesia)
SLCDA	Sri Lanka Coconut Development Authority
SPC	South Pacific Commission
TAC	Technical Advisory Committee of the CGIAR
UNCTAD	United Nations Conference on Trade and Development
UNDP	United Nations Development Programme
UNIDO	United Nations Industrial Development Organization
UPLB	University of the Philippines at Los Banos
VISCA	Visayas State College of Agriculture (Philippines)

Chapter one:

Executive Summary

Coconut is the oil crop most in need of international research
support. International research on the crop is currently underfunded and it has the potential for high pay-off.
(*CGIAR Priorities and Future Strategies, 1986*)

The coconut palm is the 'tree of life', the most important palm of the humid
tropics. This document presents a case for additional support for research, in
order to enable it to continue to play its traditional role as the essential
component of ecological systems in island and coastal communities throughout the humid tropics and to conserve the rich diversity of its genetic
resources.

The need for international cooperation on coconut research to bring
together a critical mass of expertise and resources to solve some of the
pressing problems facing the crop and its farmers, has long been recognized
amongst producing countries, coconut researchers, and development
agencies. A number of proposals have been prepared for international
support, particularly in the area of coconut germplasm collecting, evaluation
and breeding. These date back to the 1960s. All have lapsed. The needs of the
crop and the millions who depend on it for their livelihood have not abated
in the meantime.

This document outlines the priority problems which affect coconut production in many countries, identifies those problems which could be addressed through research, and sets out some new approaches by which a high
quality scientific effort could be initiated to address those problems which are
international in character, and beyond the scope of any one country to solve.

The **opportunities** for coconut are: (i) the increasing demand for oils and
fats and animal feed sources, particularly in developing countries as incomes
rise; and (ii) the ability of the coconut tree to produce a wide variety of food
and non-food products, additional to the traditional products of copra,
coconut oil and copra meal.

The **rationale** for further research and development is based on: (i) the
importance of coconut as a smallholder crop, produced largely for domestic

1

consumption: there are more than 10 million farm families (about 50 million people) involved in its cultivation; a further 30 million people in Asia alone are directly dependent on coconut and its processing for their livelihood; (ii) the increasing importance of domestic consumption of coconut in producing countries to meet the growing demand for vegetable oils; (iii) the predictions of future decreasing production in the Philippines (the world's major exporter of coconut oil), due to the increasing age of the palms; (iv) the price premiums paid for the lauric oils (coconut and palm kernel oil), primarily for their industrial uses; (v) the declining competitiveness of coconut in the marketplace, which means it is presently unable to take advantage of the expanding vegetable oil market and is losing ground to other crops, particularly oilpalm and rapeseed; and (vi) the fact that almost all the benefits from coconut research accrue to developing countries; the majority of these benefits go to the smallholder producers, and the balance to consumers in developing countries.

The **constraints** to coconut production worldwide are: (i) the low productivity of many trees, due to their increasing age, poor nutrition, and poor light interception: the world average yield of 0.5 tonnes/ha/year (copra equivalent) has not increased over the past 25 years; trees more than 30 years of age have entered into a phase of declining yield; (ii) the failure of many replanting programmes designed to replace old trees with higher yielding hybrids or locally adapted tall types: these failures have been due largely to a lack of understanding of the reasons for smallholder practices in coconut cultivation and a lack of incentives for smallholders to replant when prices are low; and poor adaptation of some of the potentially higher yielding hybrids; (iii) fluctuating productivity in the major producing countries, due to variable climatic conditions including drought and hurricane damage; (iv) the losses due to pests and diseases including several lethal diseases, some of unknown aetiology, which are killing millions of trees each year, and precluding new plantings in some areas; and (v) inefficient handling and processing, with a low farm-gate price to smallholders for a product of low quality with little added value. Some of these constraints can be addressed by research to increase the productivity both of the crop and of the coconut-growing lands.

The **main research needs** are to: (i) conserve the genetic resources of coconut, and select and propagate higher yielding varieties, which are locally-adapted, and drought- and disease-tolerant; (ii) control the major pests and diseases, especially the lethal diseases; (iii) increase the productivity of existing plantings by encouraging better agronomic practices, including improved nutrition and more productive and sustainable intercropping systems; (iv) develop more efficient means of handling and processing; and (v) diversify the products derived from coconut and actively promote new value-added products, so as to utilize fully the potential of the crop.

Successful research in these areas would improve the economic competitiveness of coconut relative to oilpalm and rapeseed. Substantial research efforts on these crops have led to production increases of approximately 10% per year for oilpalm and 7% for rapeseed, over the past two decades. In contrast, coconut production has been increasing by only about 2% per year, as a result of an increase in the area planted, but not in yield.

The coconut palm is believed to have originated in the Western Pacific, but as there are no truly wild species or sub-species, its centre of origin remains a matter of conjecture. It is now a pan-tropical crop, grown on approximately 11.6 million ha in at least 86 countries. The main producers are (i) the Philippines; (ii) Indonesia; (iii) India; (iv) Papua New Guinea and the Pacific Islands; and (v) Sri Lanka. Total annual world production is approximately 8.4 million metric tonnes of copra equivalent. Approximately 85% of production comes from Asia (15 countries) and the Pacific (19 countries). Coconut is also a locally important crop in at least 30 countries in the Americas, and in 22 countries along the coasts of East and West Africa, and the islands of the Indian Ocean.

Coconut is predominantly a smallholder crop, with at least 96% of total world production coming from smallholdings of 0.5–4.0 ha. It is an ecologically sound crop. It is able to grow in harsh environments affected by high salinity, drought, soils of low fertility and on atolls where few other plant species thrive. It plays an important role in the sustainability of often fragile ecosystems in island and coastal communities. Coconut is used as a source of food, drink, fuel, stock feed, fibre and shelter for village communities, where it is often referred to as the 'Tree of Life'. It is also a cash crop, used to produce many items for sale, at local, national, or international level. About 70% of the total crop is consumed in producing countries. Coconut is also an important export crop for some countries. The main internationally-traded products are copra, coconut oil, copra meal, desiccated coconut, and an increasing amount of coconut milk. It is especially important as an export crop from the Philippines, and from many small island countries in the Pacific and Indian Oceans and the Caribbean, where it often provides over 50% of a nation's total export earnings.

Coconut research is presently underfunded. There are many national research programmes. With few exceptions, they are not well-supported financially nor do they have sufficient appropriately trained staff and facilities. Most suffer from a lack of continuity in funding, both from national sources, and from external agencies. Many small producing countries are not able to support a coconut research programme at all. At present there are few means by which small countries can access new technologies, such as higher yielding, disease-resistant planting material. Yet they could be active participants in and contributors to an international programme.

There are also a few regional and international research activities. The lead agency among the producing countries is the Asian and Pacific Coconut Community (APCC), which primarily supports technical meetings, information and publications, and some specific research projects.

Among international research organizations, the International Board for Plant Genetic Resources (IBPGR) has supported the collecting and conservation of coconut genetic resources. The International Council for Research in Agroforestry (ICRAF) has described the principal coconut-based farming systems throughout the tropics. The newly formed Bureau for the Development of Research on Tropical Perennial Oil Crops (BUROTROP), sponsored primarily by the European Commission, has established a small secretariat in Paris. Its aims are to provide information and documentation services on coconut and oilpalm and to foster coordination of research on both crops.

Among bilateral agencies, the French Institut de Recherches pour les Huiles et Oleagineaux (IRHO) has conducted a multicountry programme over the last 30 years, concerned principally with the production of new coconut hybrids, and their adaptation and agronomic requirements, especially nutrition, over the wide range of environments that are used for coconut.

Several other bilateral and multilateral development agencies and the development banks support a range of research projects, primarily on aspects of coconut improvement, pest and disease control, and postharvest processing. These agencies also support many projects for replanting and rehabilitation of coconut lands. Many of these costly development projects have been undertaken with an inadequate research base, and have not achieved their objectives. Most of the efforts by bilateral and multilateral agencies have been characterized by their support for individual projects of limited duration (usually 3–5 years). This has led to a stop–start approach to coconut research. The present research efforts are thus not addressing the needs of the crop internationally, nor capitalizing on the promising results from research, for the benefit of smallholders.

Coconut breeding in several countries over the past 30 years has demonstrated that hybrids are capable of yielding copra at up to 6 tonnes/ha year, under favourable conditions. Progress has also been made in the identification of the causal agents of lethal diseases of previously unknown aetiology, such as cadang-cadang and lethal yellowing disease. Recent reports suggest that clonal propagation of coconut *in vitro* has been achieved with a continuous output of plantlets possible. These promising results from only a few programmes suggest that a well-organized and adequately funded international research effort could yield a high return on investment. The new technologies developed must meet the real needs of smallholders, if the return on research investment is to be realized.

The long-term nature of coconut research, the history of discontinuity and lack of adequate support in its funding, the prospects of a high return from research investment, and the likely distribution of research benefits to smallholders, make coconut a particularly suitable target for international support. This fact was first recognized by the CGIAR in the 1986 report on 'CGIAR Priorities and Future Strategies'. The CGIAR Technical Advisory

Committee (TAC) concluded that coconut was one of three priority areas requiring international support. The report states:

> Coconut is the oil crop most in need of international research support. International research on the crop is currently underfunded and it has the potential for high pay-off. Furthermore, coconut is a smallholder crop that is ecologically sound and offers a broad range of dietary, income, and employment opportunities. It is not only a primary source of edible oil, but also of fibre and livestock feed, once it has been processed into a variety of end-products. Furthermore, there appears to be good research potential for coconut . . . . TAC therefore, encourages the creation of a research network to strengthen and coordinate coconut research and supports CG system involvement in such a network.

At the CGIAR meeting in Ottawa in 1986, the Consultative Group requested TAC to give further consideration on how best to provide international support for coconut research. Subsequently, consideration of coconut research has been included within TAC's broader view of expanding the scope of the CGIAR system to include forestry and agroforestry, as well as several other existing international agricultural research centres. TAC requested the Australian Centre for International Agricultural Research (ACIAR) to assist it, by undertaking consultations with interested parties in coconut-producing countries and various research and development organizations on the needs and opportunities for research. Between 1988 and 1990, TAC considered three reports prepared by ACIAR which addressed the need and feasibility of establishing an international research initiative on coconut.

The present document synthesizes the reports prepared by ACIAR for TAC, and TAC's own deliberations on coconut research. It also makes recommendations for future action. The document discusses: (i) the current status and future trends for coconut within the context of the world fats and oils market; (ii) the importance of coconut as a smallholder crop that is an important component of long-term agroforestry systems in coastal and island regions throughout the world; (iii) the present constraints of the crop; (iv) current research efforts, at the national, regional and international level; (v) the priority problems requiring an international approach; (vi) the goals, organization and functions of an international research initiative; (vii) the possible institutional options by which an international effort might be implemented; and (viii) the next steps required.

Various options for providing additional support for coconut research are considered. These include: (i) additional/bilateral support for national programmes; (ii) an international coconut research centre; (iii) an international coconut research network; and (iv) an international coconut research council and/or (v) a coconut research consortium.

The institutional arrangements whereby an international research initiative on coconut could be implemented are also compared. These options include those which could be incorporated within the CGIAR system and others which could be conducted under international auspices but outside the CGIAR system.

In the light of its consideration of coconut research, TAC recommended to the CGIAR in 1990 that coconut be included in the CGIAR portfolio of activities. TAC further recommended that the research areas which warranted an international effort were in the fields of: (i) germplasm collecting, conservation, evaluation and enhancement; (ii) the control of diseases and pests, especially the lethal diseases; (iii) the productivity and sustainability of coconut-based agroforestry systems; (iv) the need for greater efficiency and added value in postharvest handling and utilization; and (v) socioeconomic issues, especially the factors which influence farmers' participation in rehabilitation and replanting.

With regard to the possible institutional arrangements, TAC saw a clear need for networks with a strong 'enabling' component to fund research. TAC considered that provision for this enabling component should be made through close cooperation with an existing development agency. Further, TAC stressed that close cooperation with IBPGR would be required with respect to coconut germplasm collecting and conservation. TAC also recommended that IBPGR be invited to establish and manage a small germplasm research unit for coconut in the Asia/Pacific region.

In the 1992 report on CGIAR priorities and strategies, TAC reiterated its earlier recommendation for the inclusion of coconut within the CGIAR portfolio. The recommendations of this report were adopted by the CGIAR at its meeting in Istanbul in May 1992, whereby the CGIAR agreed to the inclusion of coconut within its portfolio of activities, and to the five priority areas for international support, as recommended by TAC. It also agreed that genetic resources should be given first priority, and encouraged IBPGR to strengthen its work on coconut genetic resources.

In regard to the means of implementation of a broader international initiative on coconut research, which addressed the five priority areas identified by TAC, the CGIAR recommended that the question of institutional mechanisms be taken up in the context of the CGIAR's expanding support for agroforestry and forestry, and the inclusion of ICRAF and the newly established Centre for International Forestry Research (CIFOR) in the CGIAR system.

There are several options by which an international research programme on coconut could be implemented, either within the CGIAR system, or under international auspices but outside the CGIAR system. In part, the scope of such an initiative will depend on the financial resources available, and the interest either of the management and Boards of Trustees of prospective host institutions in expanding their coverage to include coconut, or of prospective sponsors establishing an independent initiative outside the CGIAR system.

The critical elements of an international initiative are (i) to identify a set of priority problems of global significance; (ii) to provide an 'enabling mechanism' by which research on the priority problems could be undertaken on a contractual basis by scientists in national programmes, regional or-ganizations and/or laboratories in industrial countries; (iii) to provide inter-national auspices for a programme of coconut genetic resource collecting, conservation, exchange, and utilization; and (iv) to provide a mechanism for continuity of funding for coconut research.

In the light of these considerations, it is recommended that the next steps are as follows:

1. **To establish an international coconut genetic resources network under the auspices of IBPGR.**
   An international coconut genetic resources network is being fostered following the plan of action developed by participants at an international workshop on coconut genetic resources held in Cipanis, Indonesia in October 1991.

2. **To establish a Coconut Research Consortium.**
   An important component of any international initiative will be an 'enabling mechanism', by which research can be commissioned on an agreed set of priority problems of importance to many producing coun-tries but which are beyond the scope of any one country or agency to support. Thus, it is proposed that research within the priority areas which is international in character could be supported through a Coconut Research Consortium, established to provide this enabling mechanism.

The initial tasks of such a body as the proposed Coconut Research Consor-tium would be to: (i) support the establishment of an international coconut genetic resources network; and (ii) mobilize additional financial and technical support to commission (possibly on a competitive basis) research on the set of priority problems of global significance as identified by TAC.

The consortium could be comprised of representatives of producers and consumers, including the importers of coconut products, interested bilateral and multilateral development agencies, and other organizations active in their support of coconut research. The Coconut Research Consortium could be advised by a technical committee knowledgeable about the crop, and the research needed to underpin its development. It is proposed that one of the member agencies of the Consortium provide the necessary secretariat services.

A case is presented for establishing an international research initiative on coconut. This subject has been examined since the early 1970s by several bodies interested in improving the productivity of coconut, and increasing the income of millions of smallholders dependent on the crop. Although the problems of the crop have been identified, and the potential returns from research appreciated, all these efforts have lapsed. The key problem has been the lack of follow-through to establish a consortium of producing countries, importers, and development agencies, that would design and implement a high quality research programme that addresses the major issues facing the crop globally, and provides the continuity of funding that is essential for a perennial tree crop such as coconut. As the next step, a consortium for coconut research needs to be established, in order to translate recommendations into reality.

Chapter two:

Origin and Importance of the Coconut Palm

'Coconut has been present in the Pacific Islands for millions of years ...'

Origin and Distribution

Centre of origin

Coconut, *Cocos nucifera L.*, is the most important palm of the wet tropics. It has a pan-tropical distribution, occurring in coastal areas between 20°N and 20°S of the equator (Fig. 2.1). The coconut has a long history in both the eastern and western hemispheres and its widespread and apparently ancient occurrence in both has led to uncertainty as to its centre of origin. It is believed that although the wild ancestor of coconut may have been from South America, the wild type was dispersed widely millions of years before the coconut was domesticated in the Indo–Pacific area (Harries, 1992a,b).

There is convincing evidence that the coconut was distributed widely by the nuts floating in ocean currents and germinating after they were washed ashore in new locations; and also that it was carried by man as a source of food and drink on long sea voyages. Coconut has been present on most of the Pacific Islands for millions of years, long before their settlement by the Polynesians. Coconut has a recorded history of some 2000 to 3000 years in coastal areas of Sri Lanka and southern India and of less than 500 years in West Africa and the Americas (Purseglove, 1975; Harries, 1978).

Importance

Coconut is an integral part of the functioning of local communities. It is often referred to as the '*tree of life*', since almost every part is used to make some item of value to the village community (Fig. 2.2). The importance of coconut

9

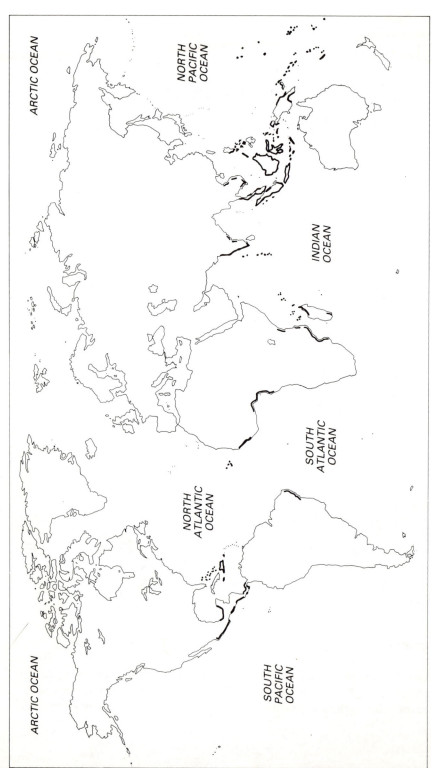

Fig. 2.1. Coconut-growing regions; darkened coastlines indicate regions. Source: Plucknett (1979).

Fig. 2.2. Coconut: the tree of life, and its end-products. Redrawn from UCAP, 1981.

is thus closely associated with its suitability in providing a supply of food, drink and shelter at the village level, as well as copra, coconut oil and other products for local cash sale and export earnings.

In subsistence and semi-subsistence farming systems, coconut provides a reliable insurance as a supply of food even when other crops fail, and a source of cash. Coconuts are an important component of the daily diet, and a significant contributor to human nutrition, providing a source of energy and vitamins.

Purseglove (1975), Harries (1978, 1992a,b) and Ohler (1984) provide a brief history of the development of coconut as a subsistence and an export crop. Although coconut has been grown as a subsistence crop for thousands of years, it became important as an export crop only in the 1840s. The industrial process for making soap required a cheap source of oil. Coconut oil derived from copra (the dried endosperm of the nut) provided that source. The development of dynamite from nitroglycerine between 1846 and 1867 turned glycerine, a once discarded by-product of soap manufacture, into another profitable product. Coconut oil also replaced animal fat in the manufacture of margarine. The various uses of coconut are described in more detail in Box 2A and illustrated in Figs 2.2, 2.3 and Table 2.1.

About 70% of the coconut crop is consumed in the producing countries, while 30% is traded internationally. Coconut is consumed daily in the common diet of the inhabitants of the producing areas. There are more than 100 products made directly or indirectly from coconut. These vary from simple cooking utensils used in the village, to high value-added products such as coco-chemicals and activated charcoal. The most important products in world trade are copra, coconut oil, copra meal, desiccated coconut, coir fibre, shell charcoal and an increasing amount of coconut milk. The coco-chemicals which are becoming increasingly important and valuable are methylesters, fatty alcohols, and glycerine.

In villages and towns, the white coconut meat or the coconut milk extracted from it, provide the basis for many dishes, as described elsewhere (Persley *et al.*, 1990). Coconut oil is used for cooking, lighting and lubrication, and for the manufacture of margarine, bakery products, fats, soaps, detergents and toiletries. Coconut meal (the residue left after the oil is removed from the copra) is used for animal feed. The timber can be used for load-bearing structures in buildings and for the manufacture of furniture, while the leaves are plaited for roofing material. The husk is used to make fibres for ropes and matting. The shell is heated to make charcoal as a local fuel source. Various artifacts are made from the leaf and the shell.

Nutritional composition of coconut oil

The chemical composition of coconut oil is 91% saturated and 9% unsaturated fatty acids. The melting point is about 24–27°C (Ohler, 1984).

USES OF COCONUT OIL AND ITS DERIVATIVES

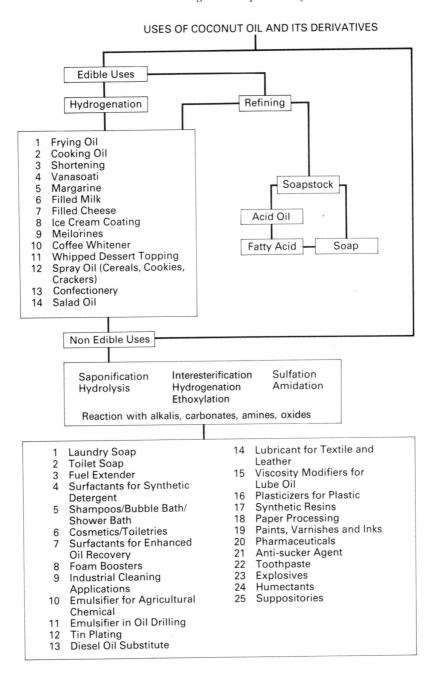

Fig. 2.3. Uses of coconut oil and its derivatives. Source: Philippines Coconut Authority.

Table 2.1. Non-edible uses of coconut oil.

Chemical products	End uses
Esters	Surfactants, plasticizers, cosmetics, lubricants, diesel oil substitute, anti-sucker agent
Fatty alcohols and derivatives	Surfactants, non-ionic detergents, solubilizers, softeners, dispersant defoamers, cosmetics
Fatty acid salts	Soaps
Short carbon chain	Plasticizers, lubricants, coatings, cosmetics, perfumes, flavours, adhesives
Amines	Dyes, pharmaceuticals, plasticizers, emulsifiers, detergents, fabric softener, lubricants
Amides	Adhesives, resins, inks, surfactants, foam stabilizer
Glycerine	Toothpaste, suppositories, humectant, pharmaceuticals, explosives

Source: Spectrum of Coconut Products, Philippines Coconut Authority (PCA, 1986)

Like its competitors palm kernel oil and babassu oil, it is a lauric oil, because of the high percentage of lauric acid it contains (in the range 44–51%, depending on the variety and the environment) (Ohler, 1984).

There has been debate over the past years as to the health risk posed by 'tropical oils' (e.g. coconut oil, palm kernel oil and palm oil). Some researchers have argued that because of its high proportion of saturated fatty acids, coconut oil contributes to the build-up of cholesterol levels in humans (Jones, 1989). Research at the Harvard Medical School (1987) gives substantial support to the view that there is no scientific basis for describing coconut oil as a health risk. The testimony of the Harvard scientists to the US Senate is given in Box 2B.

Saturated fats have sub-groups known as medium-chain and long-chain fatty acids: medium-chain fatty acids do not raise cholesterol levels, but long-chain fatty acids may do so. Long-chain polyunsaturates also form cholesterol esters, which may be deposited in the arteries.

The most commonly used fats and oils classified as medium chain are coconut oil and palm kernel oil. Those classified as long-chain saturated fats and oils are cocoa butter, palm oil, dairy fats, lard, tallow and stearines (Enig, 1990).

It appears that it is the chain length which is the important factor in identifying the risk posed by certain fats, not the amount of saturated fatty acids *per se* (Enig, 1990). Although coconut contains a high proportion of saturated fatty acids, these are short-chain fatty acids which are quickly burned off as a source of energy, and which are not available for incorporation into body fat, nor for synthesis of cholesterol. Indeed, for many countries the problem is to provide the increasing population with sufficient vegetable oil in the diet, rather than its over-consumption.

Research continues by nutritionists to demonstrate any risks and benefits associated with the consumption of saturated and unsaturated fats and oils. At present, it appears that there is little evidence to support the view that coconut oil is a health risk at the levels it is normally consumed. Indeed, the damaging perception of tropical oils as a health risk in North America may have more to do with trade concerns, and the competition between temperate and tropical oils, than the health of consumers.

Coconut Production Statistics

Coconut is grown on approximately 11.6 million hectares spread over at least 86 countries. The area under coconut has been increasing at approximately 2% p.a. for the past several decades. Reliable statistical data are difficult to obtain. The best available source of quantitative data is the Coconut Statistical Yearbook, published annually by the Asian and Pacific Coconut Community (APCC), in Jakarta. The APCC member countries are India, Indonesia, Malaysia, Philippines, Sri Lanka, Thailand, Papua New Guinea, Solomon Islands, Vanuatu, Western Samoa, Federated States of Micronesia and Palau. The APCC statistics are derived from information supplied directly by APCC member countries, supplemented with information from FAO production yearbooks and other sources (APCC, 1990).

The major producing areas are in Asia and the Pacific, where about 85% of the crop is grown. The FAO production statistics for 84 counties are given in Table 2.2. The area under coconut in various regions is shown in Table 2.3. There are 13 producing countries in Asia, and 19 in the South Pacific. The major producers are the Philippines, Indonesia, India, Papua New Guinea and the Pacific Islands, and Sri Lanka. Coconut is also locally important in coastal areas and island communities in the Americas, which produce about 7% of the total world crop. There are some 32 producing countries in the Americas, extending from Florida, USA, to Brazil. The major producers are Mexico and Brazil. Elsewhere, it is important as a food and oil crop in the coastal areas of 22 countries in West and East Africa, and the islands of the Indian Ocean, which produce about 6% of the total world crop. The major producers in Africa are the Côte d'Ivoire, Mozambique and Tanzania.

The detailed production data for the APCC member countries are given in Table 2.4. The figures show that coconut is predominantly a smallholder crop produced for local consumption. Almost all production (96%) comes from smallholdings of 0.2–4 ha. The average size of holdings in some areas, such as southern India, is only 0.2 ha (World Bank, 1991). Approximately 70% of total production in Asia is consumed in the producing countries. The

Table 2.2. FAO coconut production statistics, 1979–88.

	Coconut production ('000 tonnes)				Copra production ('000 tonnes)			
	1979–81	1986	1987	1988	1979–81	1986	1987	1988
World	34750	38538	37649	36802	4437	5466	4914	4513
Africa	1555	1741	1799	1879	180	220	231	220
Benin	20	20	20	20	3	3	3	3
Cameroon	3	4	4	4	1	1	1	1
Cape Verde	10	10	10	10				
Comoros	53	53	50	53	3	3	3	3
Côte d'Ivoire	156	406	450	500	21	61	70	56
Eq. Guinea	7	8	8	8				
Ghana	160	110	110	112	7	7	7	7
Guinea	15	15	15	15	3	3	3	3
Guin Bissau	25	25	25	25	5	5	5	5
Kenya	85	71	72	73	6	11	12	12
Liberia	7	7	7	7				
Madagascar	67	79	79	82	10	9	9	9
Mauritius	5	3	3	3				
Mozambique	453	410	415	420	72	67	68	69
Nigeria	90	100	100	105	10	12	12	13
Sao Tome Prn	36	35	35	37	4	4	4	4
Senegal	4	5	5	5				
Seychelles	22	20	21	19	4	3	4	3
Sierra Leone	3	3	3	3				
Somalia	10	14	15	15				
Tanzania	310	330	340	350	29	29	29	29
Togo	14	14	14	14	2	2	2	2
N C America	1412	1689	1649	1678	187	239	214	204
Barbados	2	2	2	2				
Belize	3	3	3	4				
Costa Rica	25	27	27	28	2	2	2	2
Cuba	17	22	24	24				
Dominica	10	14	14	14	1	2	2	2
Dominican RP	69	73	71	129	6	15	14	26
El Salvador	55	82	75	78	3	3	3	3
Grenada	7	8	8	8		1	1	1
Guadeloupe	3	3	3	3				
Guatemala	2	2	2	2				
Haiti	34	37	37	38				
Honduras	16	6	8	9	2	1		1
Jamaica	176	178	180	180	7	8	8	8
Martinique	1	3	3	3				
Mexico	851	1079	1042	1006	149	196	172	150

Table 2.2.   Continued.

	Coconut production ('000 tonnes)				Copra production ('000 tonnes)			
	1979–81	1986	1987	1988	1979–81	1986	1987	1988
Nicaragua	1	3	3	3				
Panama	23	21	20	22	1			
Puerto Rico	9	7	7	6	2	1	1	1
St Kitts Nevis	2	2	2	2				
St Lucia	34	31	31	31	6	5	5	5
St Vincent	22	25	24	25	2	1	1	1
Trinidad Tob	49	62	62	62	5	5	5	5
South America	855	947	977	1029	38	34	36	36
Argentina								
Brazil	507	588	599	669	2	3	2	3
Colombia	64	88	88	75				
Ecuador	73	40	42	35	13	7	7	5
Guyana	30	41	43	45	3	3	4	5
Peru	15	15	15	15				
Suriname	6	8	11	11	1	1	2	2
Venezuala	160	167	179	179	19	20	21	21
Asia	28585	31817	30906	29900	3697	4617	4135	3757
Bangladesh	75	83	83	86	3	3	3	3
Burma	95	250	321	330				
China	59	75	80	83				
India	4192	4759	4458	4600	374	350	330	350
Indonesia	11333	11650	11600	11500	1049	1150	1270	1250
Kampuchea DM	28	41	42	44	5	7	8	8
Malaysia	1121	1187	1167	1186	141	121	118	120
Maldives	7	12	12	12	1	2	2	2
Philippines	8811	9451	9443	8640	1897	2564	2100	1700
Singapore	10	5	5	5	4	8	8	8
Sri Lanka	1692	2310	1742	1440	128	243	125	65
Thailand	781	1280	1350	1350	38	58	61	65
Vietnam	290	625	605	625	58	112	109	115
Oceania	2344	2345	2318	2315	334	336	298	296
Amer. Samoa	5	5	5	5				
Cocos Islands	4	6	6	6	1	1	1	1
Cook Islands	13	10	10	10	1	1		1
Fiji	217	226	150	152	22	22	13	13
Fr. Polynesia	132	150	150	150	17	14	15	15
Guam	31	34	36	37	1	2	2	2
Kiribati	76	50	90	90	9	6	6	12
Nauru	2	2	2	2				

Table 2.2. Continued.

	Coconut production ('000 tonnes)				Copra production ('000 tonnes)			
	1979–81	1986	1987	1988	1979–81	1986	1987	1988
New Caledonia	14	11	14	14	1	1	1	1
Niue	3	2	2	2				
Pacific Islands	211	195	193	195	29	25	25	25
Papua New Guinea	835	820	900	910	145	158	149	155
Samoa	171	210	200	200	20	25	18	14
Solomon Islands	201	209	208	200	32	32	27	25
Tokelau	2	2	2	3				
Tonga	95	68	45	53	12	8	5	6
Tuvalu	5	3	3	3				
Vanuatu	326	340	300	281	43	42	36	26
Wallis etc	3	3	3	3				
Developed M E								71
Other Developed								71
Developing M E	34373	37797	36922	36050	4373	5327	4798	4320
Africa	1555	1741	1799	1879	180	220	231	220
Latin America	2267	2636	2625	2708	225	273	250	240
Far East	28208	31076	30179	29149	3634	4498	4018	3563
Other Developing	2344	2345	2318	2315	334	336	298	296
Central Planned	377	741	727	752	63	119	117	123
Asian CPE	377	741	727	752	63	119	117	123
Developed All								71
Developing All	34750	38538	37649	36802	4437	5446	4914	4442

Source: FAO Production Yearbooks (1969–1988)

major exporter is the Philippines. In the Pacific Islands, with their smaller populations, about 30% is consumed locally and 70% exported.

Estimates by the World Bank (1991) also indicate that well over half the total crop is consumed in the producing countries. The World Bank predicts that although total coconut production is expected to increase (from 2.9 million tonnes of oil equivalent in 1985 to 3.7 million tonnes by 2000), exports are expected to increase only from 1.26 million tonnes in 1985 to 1.35 million tonnes in 2000. Much of the expected increase in production will be used to meet the growing demand for vegetable oils in producing countries.

The aggregate figures conceal much wider variations among coconut-producing countries. These vary from countries such as Indonesia and India which are large producers with large populations, which usually consume almost all their crop domestically, through to the many small island countries which export most of their crop. For the islands of the Caribbean, Indian and

Table 2.3. Area of coconut worldwide, 1985–89 ('000 ha).

Country	1985	1986	1987	1988	1989
APCC Countries	**9358**	**9479**	**9538**	**9750**	**9861**
F.S. Micronesia	17	17	17	17	17
Fiji	21	23	23	23	23
India	1183	1209	1231	1346	1473
Indonesia	3050	3113	3153	3241	3317
Malaysia	315	299	294	289	286
Papua New Guinea	241	241	241	260	260
Philippines	3270	3284	3252	3222	3110
Solomon Islands	59	59	59	59	59
Sri Lanka	419	419	419	419	419
Thailand	415	414	414	407	407
Vanuatu	93	94	94	95	96
Vietnam	214	245	277	310	333
Western Samoa	47	48	50	48	47
Palau	14	14	14	14	14
Other countries	**693**	**701**	**720**	**720[1]**	**720[1]**
Africa	371	383	396		
Ivory Coast	32	33	34		
Tanzania	260	270	280		
Others	79	80	82		
N.C. America	**173**	**168**	**168**		
Mexico	110	105	105		
Jamaica	50	50	50		
Others	13	13	13		
S. America	**25**	**27**	**27**		
Asia	**66**	**67**	**70**		
Bangladesh	31	31	32		
Burma	25	26	28		
Others	10	10	10		
Pacific	**58**	**56**	**59**		
Kiribati	36	33	36		
Others	22	23	23		
Total	**10051**	**10180**	**10258**	**10470**	**10581**

Source: APCC (1990)
[1] Data refer to 1987

the Pacific Oceans, coconut is often both their primary subsistence crop, and their only significant source of foreign exchange. There are few, if any, alternative crops to serve the needs of these many small countries.

Estimates by the APCC indicate that there are more than 10 million farm

Table 2.4. Coconut production data for Asian–Pacific coconut community area: producers, domestic production, consumption.

Country	Area ('000 ha)	Ave. size holding (ha)	No. farm families ('000)	Estimated % prod. from small-holdings	Total coconut production		Estimated domestic consumption		% Crop consumed domestically	Exports							Coconut export earnings US $ millions	% Coconut contribution to country exports
					Nut equiv million nuts	Copra equiv '000 tonnes	Nut equiv million nuts	Copra equiv '000 tonnes		Copra '000 tonnes	Coconut oil '000 tonnes	Copra meal '000 tonnes	Desic coconut '000 tonnes	Shell '000 tonnes	Fibre '000 tonnes	Other '000 tonnes		
Asia-APCC																		
India	1209	2	5000	98	6620	988	6743	1006	100	–	–	–	–	–	24	–	26	0.3
Indonesia	3182	4	3000	97	10491	2098	10397	2079	99	–	–	380	303	1	–	–	37	0.25
Malaysia	315	3.5	90	99	1054	200	793	150	75	41	49	–	7	–	–	–	26	2.2
Philippines	3262	3	1000	98	11675	2631	2106	402	15	136	1238	818	68	27	n/a	129	558	11.5
Sri Lanka	419	0.5	715	99	3040	617	1916	391	63	10	85	40	60	35	85	–	101	8.3
Thailand	415	n/a	n/a	99	707	200	541	153	76	–	–	6	–	–	5	415	3	0.04
Total APCC-Asia	8802	Range 1.5–3.5 ha	9805 +	98*	33587	6734	22506	4781	70*	187	1372	1244	438	63	114	544	751	
Pacific-APCC																		
Papua New Guinea	241	1.5	104	60	1248	208	300	50	24	93	41	20	–	–	–	–	23	2.2
Solomon Islands	63	4*	60*	80	209	46	81	18	38	32	–	–	–	–	–	–	3	5.2
Vanuatu	93	4*	50*	70	429	59	131	18	30	40	–	–	–	–	–	–	4	50
Western Samoa	48	n/a	n/a	90	159	35	46	10	28	3	14	6	–	–	–	–	4	35
F.S. Micronesia	16	15	n/a	99	50	10	49	9	99	1	–	–	–	–	–	–	n/a	n/a
Total APCC-Pacific	461	Range 1.5–4.0 ha	214 +	62*	2095	358	607	105	30*	169	55	26	–	–	–	–	34	–
Total APCC Asia/Pacific	9263	Range 0.5–4.0 ha	10019 +	96	35682	7092	23113	4886	69*	356	1427	1270	438	63	114	544	785	

Source: APCC 1986
*approximately

families (approximately 50 million people) directly involved in coconut cultivation (APCC, 1986). A further 30 million people in Asia alone are dependent directly on coconut and its processing for their livelihood. The social significance of coconut is such that all family members are involved in its cultivation, harvesting and local processing. Women play an important role in copra making in the villages, and children often gather the nuts for copra making. It is also important in poverty alleviation as it is a source of ready cash income for families.

Many people are concerned with the further processing of copra into coconut oil and other products in producing countries. Figures for the proportion of the labour force involved in coconut-based processing activities are difficult to obtain. However, in the Philippines, it has been established that about one-third of the population are directly dependent on coconut for their livelihood. In Sri Lanka the coconut industry is also a major employer, and coconut is second only to rice in importance as an agricultural crop.

The role of coconut in selected producing countries is described in the following section.

The Role of Coconut in the Economy

Asia

Philippines

Coconut is one of the major components of the economy in the Philippines. Coconut is the main earner of foreign exchange, contributing about 13% of total annual export earnings. It has been estimated that of 10 million ha of agricultural land in the Philippines, about one-third is planted to coconut (Ohler, 1984). The area under cultivation (approximately 3.3 million ha) is second only to rice. There are about one million farm families growing coconut. The Philippines is the world's largest producer of coconut and the major exporter of coconut products.

Indonesia

Indonesia is the world's second largest coconut grower, producing almost as much as the Philippines. The total coconut area is approximately 3 million ha, of which 97% is on smallholdings of less than 4 ha (APCC, 1986). There are approximately 3 million farm families growing coconut. Average yield is only 0.5 tonnes/ha year. Most of the crop is used for local consumption. Coconut cultivation plays an important role in the economic and social life in Indonesia, where its status ranks second after rice. It is estimated that the demand for coconut will soon exceed supply in Indonesia, with demand rising to as much as 1.0 million tonnes per annum. There is thus a pressing need for an effective replanting programme, and for an increase in yield from the existing plantings.

India

Coconut has a documented history in India of some 3000 years. It is the world's third largest coconut producer, with approximately 1.2 million ha under cultivation, principally in four southern states (Kerala, Tamil Nadu, Karnataka and Andhra Pradesh). The total area under coconut has been increasing by about 3% per year, but average yield appears to be declining. Almost all production is consumed domestically. India is also an importer of coconut and other vegetable oils, and demand is increasing. Almost all production comes from smallholdings of 2 ha or less. There are approximately 5 million farm families growing coconuts. It is estimated that a further 10 million people in India depend directly or indirectly on coconut for their livelihood through processing and sale of the crop (APCC, 1986).

Sri Lanka

Sri Lanka has approximately 0.4 million ha under coconut, which ranks second after lowland rice in land use. About 28% of agricultural land is planted to coconut (Ohler, 1984). Approximately 99% of the coconut area is made up of 715,000 smallholdings. The average farm size is 0.5 ha. The crop also provides employment for about 5% of the nation's workforce (APCC, 1986).

Approximately 40% of the Sri Lankan crop is consumed as fresh nuts. The balance (60%) is processed into copra and desiccated coconut. Copra is further processed to coconut oil for local use and for export. Coconut shell and fibre are also processed in small or medium scale industrial units (APCC, 1986).

Coconut exports are important sources of foreign exchange for Sri Lanka, comprising approximately 8% of total export earnings in 1986; it is also an important component of the diet, providing about 20% of the daily calories for the rural population. Domestic *per capita* consumption has been increasing by about 1.8% per year. Unless production can keep up with domestic consumption increases, export volumes will fall and Sri Lanka will lose a valuable source of foreign exchange.

In 1984, the Sri Lankan Government outlined a coconut development strategy for the next decade. It was designed to raise production at a rate sufficient to meet domestic demand and maintain exports. It was concluded that production should be increased from 2.4 to 3.0 billion nuts per year. This increase is proposed to come from the greater use of fertilizers (provided at subsidized prices), grants to smallholders to encourage coconut rehabilitation, improved extension services, and better access to credit for smallholders to encourage them to move to a higher input system (IRHO, 1986).

Sri Lankan production tends to fluctuate, largely due to climatic effects. With production in 1986 at a level of approximately 3 billion nuts, domestic needs were met, and export volume was increased. In 1987, production decreased, largely due to drought. Production trends over the next few years will be important to determine if the government policies are taking effect.

Malaysia

Although in recent years oilpalm has replaced coconut as the major export oil crop from Malaysia, coconut still ranks fourth in terms of land utilization. There are approximately 315,000 ha under cultivation. Production was about 200,000 tonnes of copra equivalent in 1986. Coconut is primarily a smallholder crop in Malaysia, although some plantations use coconut as a shade crop for cocoa (APCC, 1986).

Thailand

Thailand is a small coconut producer in Asia, relative to Indonesia and the Philippines. In 1986, production was 200,000 tonnes of copra equivalent, harvested from approximately 415,000 ha, mainly in southern Thailand. Thailand has been attempting to increase coconut production by the use of hybrids and to advocate to smallholders the use of coconut as an intercrop or shade crop. Thailand produces sufficient to meet its own needs for coconut products, but rarely exports them.

Papua New Guinea and the Pacific Islands

Coconut is an indigenous species in the Pacific Islands and may have first reached many of the islands by natural dispersal millions of years ago. The region is rich in genetic diversity, with nut types ranging from the Niu Vai of Samoa (the largest known coconut fruit) to the small-fruited types of Tuvalu. The genetic diversity in the region has been described by Whitehead (1966) and Foale (1987).

Coconut is the most important crop in the South Pacific. It is the dominant subsistence crop for village people, particularly on atolls and outer islands. It is often the only substantial source of foreign exchange for a nation. The region is collectively the world's fourth largest producer after the Philippines, Indonesia and India. Approximately half the region's production comes from Papua New Guinea, and the remainder from another 18 Pacific Island countries. A major problem throughout the region is the increasing senility of palms. For example, more than half the palms in Papua

New Guinea are beyond their economically productive life. A similar situation exists in most other Pacific countries.

The present systems of harvesting and processing coconuts are labour-intensive and wasteful of potentially useful by-products. More self-contained units, capable of providing their own fuel from combustion of coconut, and of making better use of the total coconut resource would be especially valuable to island communities. Etherington (1988) has described the potential for such systems as part of an overall strategy to make coconut an even larger contributor to subsistence affluence in the South Pacific.

Papua New Guinea

Coconut is an important crop in Papua New Guinea, although not as dominant as in the early 1920s when it provided 90% of total export earnings. There are approximately 104,000 farm families cultivating coconut; this represents about half the country's rural population (APCC, 1986). There are approximately 240,000 ha under cultivation. Production in 1986 was some 200,000 tonnes (copra equivalent) of which 24% was consumed locally, and 76% exported as copra, coconut oil and copra meal. Papua New Guinea is the second largest exporter of copra on the world market.

Solomon Islands

There are approximately 63,000 ha of coconut in the Solomon Islands, of which 49,000 ha are farmed by smallholders. Intercropping of food and cash crops is widely practised by smallholders. About 25% of production is consumed locally, and 75% exported as copra. Coconut provides about 5% of total export earnings.

Vanuatu

Coconut is widely grown in the coastal areas of Vanuatu, growing on approximately 93,000 ha. Approximately two-thirds of this area is cultivated by smallholders. Coconut provides employment and income for most of the rural population. Approximately 30% of production is consumed locally and the balance exported as copra. Export earnings from coconut represent about half of all export earnings.

Western Samoa

Western Samoa has approximately 48,000 ha under coconut cultivation. This represents about 34% of the arable land. Approximately 90% of the coconut area is cultivated by smallholders, and the balance by a government-owned plantation.

Coconut producers are the main earners of foreign exchange, and provide about 35% of total export earnings. Approximately 30% of the crop is consumed locally, and the balance exported, mainly as coconut oil and coconut meal.

Other Pacific Islands

Coconut is grown in many other small nations, dependencies, and states, throughout the North and South Pacific, such as the Cook Islands, French Polynesia, Guam, Hawaii, Kiribati and Tuvalu.

The Americas

The coconut-growing areas in the Americas extend from Florida to Brazil, with many countries (32) each having a small industry. Production is modest relative to the major producing countries in Asia, but coconut is an important crop locally, primarily for domestic consumption, and for a small export industry.

In the more temperate parts of the region (such as Florida, Bermuda and the Bahamas), the importance of coconut lies more in its role as an attractive feature of the landscape and as an ornamental palm for the horticulture industry.

Mexico

Mexico is the most important coconut producer in the Americas, accounting for approximately 60% of the region's production. All is used for local consumption. Indeed, Mexico is a net importer of vegetable oils, and proposes to plant an additional 350,000 ha of coconut to contribute toward self-sufficiency. The finding of lethal yellowing disease in Mexico in 1982 is a serious threat to the planned expansion of the industry, as the disease is spreading rapidly.

Brazil

Coconut is widely grown along the coastal areas of north eastern Brazil, where it was introduced by the Portuguese in the sixteenth century. Over 60% of Brazil's total production of approximately 700,000 tonnes is from the northern coastal strip, about 50 km wide, in the states of Alagoas, Sergipe and Bahia. On average, a coconut palm yields 25–30 nuts per year in Brazil. Green coconuts are harvested for direct consumption of coconut water as a beverage. Mature nuts support an industry which produces coconut milk and grated coconut; copra is not made. Locally the palm leaves are utilized for thatching and fencing. Brazil does not export coconut products.

The area under coconuts has also undergone steady expansion. From 73,583 ha in 1960, it increased to 117,193 ha in 1970 and to 164,770 ha in 1980. Currently the Brazilian government has a major programme underway to encourage the establishment of new large-scale plantings of coconut. Part of the programme is to make available improved coconut varieties, including dwarfs, for the new plantings as well as for the needed replacement of older plantings (Johnson and Nair, 1989).

Africa

In February 1991, the Tanzanian National Coconut Development Programme (NCDP) and BUROTROP sponsored an African Coconut Seminar. Its purpose was to review the current status of coconut in Africa and identify future research needs. A brief report on the meeting has been published by BUROTROP (1991a). At the seminar, participants from various African countries described the present national situation for coconut, in terms of its agriculture, trade, processing, research, and future priorities. A synopsis of the importance of coconut in selected African countries, drawn primarily from these country statements is given below.

Benin

Coconut is grown in Benin on 10,000 ha along the West African coast. Coconut is used primarily for domestic consumption, with a small export trade in fresh nuts developing with Togo. There is interest in a wider use of coconut in snacks, sweets, coconut milk and butter.

Cape Verde

There are approximately 3000 ha of coconut in Cape Verde. Cape Verde is the probable site of the introduction of coconut into the Western Hemisphere in 1499, and the centre of distribution to other countries of West Africa. New varieties are being introduced to increase yields.

Comoros Islands

The Comoros Islands have 30,000 ha of coconut palms. Production (50,000 tonnes copra equivalent per annum) does not meet local demand, and there are current efforts to increase production, improve oil extraction technology, and develop local processing of coconut meat and by-products.

Côte d'Ivoire

Côte d'Ivoire produces approximately 500,000 tonnes of coconut (copra equivalent) annually, making it one of the larger producers in Africa. An extensive genetic resources collection is located at the Marc Delorme Coconut Research Station. Hybrids have been bred in the Côte d'Ivoire which have a yield potential of 6 tonnes copra/ha/year (see Chapter four).

Ghana

There are approximately 36,000 ha of coconut in Ghana, mainly along the west coast. Total production is of the order of 112,000 tonnes per annum, of which 80% comes from smallholdings between 0.5–5 ha. Most of the production is used at the village level to make cooking oil. There is interest in small-scale food processing to make a variety of coconut food products.

Mozambique

Mozambique has approximately 95,000 ha of coconut palms. Production fluctuates widely, due to climatic conditions and political disturbances. In 1989, production was 35,000 tonnes compared to 13,800 tonnes in 1988. A lethal yellowing type disease is causing losses on about 1000 ha in northern Mozambique

Tanzania

Coconut provides 40–50% of national vegetable oil demand. Annual production is approximately 90,000 tonnes of copra equivalent. The yield is only 23–25 nuts per palm. This is attributed to low application of fertilizer, poor husbandry, and erratic rainfall distribution. Approximately 95% of growers are smallholders. There is an intensive government effort to increase production through the National Coconut Development Programme which is supported by the Government of Tanzania, the World Bank and Germany.

Other African and Indian Ocean Countries

Coconut is grown as a subsistence crop along the coasts of Kenya, Somalia and Oman. It is also a subsistence and an export crop for the island nations of the Malagasy Republic and the Seychelles.

The Role of Coconut in the Environment

Environmental range

Coconut is a crop of the lowland humid tropics. It grows best at altitudes below 1000 m and near the coast. It is most common and grows best on sandy coastlines, under conditions of high humidity, a mean temperature range of

25–30°C, and a loose, free-draining, and well aerated soil. Approximately 90% of production comes in the band between 20°N and 20°S latitude.

Coconut will grow in a wide range of environments. However, yield is affected by climatic factors, particularly temperature and rainfall. For optimum production, coconut requires an average temperature of 29°C, with a diurnal fluctuation not exceeding 7°C, and an annual rainfall of at least 1800 mm, evenly distributed throughout the year.

Coconut will also grow in less ideal environments. Indeed it will grow in situations such as coral atolls, which few other usable plants will tolerate. It is moderately tolerant to drought, due to its extensive root system and its ability to tolerate moderate salinity in the soil water. It can also tolerate sea spray and alkaline conditions. Its tolerance to high winds is important in areas prone to typhoons, such as the Philippines, the Pacific Islands and the Caribbean. Coconut palms will even survive occasional brief frosts in southern Florida.

Global warming

Long before the presence of humans, coconut had spread naturally to widely scattered areas where it survived unaided. It is just these areas on isolated coasts and islands which are now considered to be sensitive to the effects of global warming. The coconut is the best possible choice of a plant to prevent the degradation of the costal areas and improve the resilience of these sensitive environments. If global warming does cause the sea level to rise and coral atolls to be inundated, the coconut is the one plant species that will survive. The nuts will float in the sea, as in times past, and germinate on new shores. As new coastlines form above the present level, the coconut will be an early colonizer of the new atolls which form.

Sustainable production systems

Coconut has been planted in all possible areas of the tropics. It is found on the coast and at distances hundreds of kilometers inland and at altitudes up to 1000 m above sea level. Inevitably some sites are better than others and many sites where coconut is grown are unsuitable for worthwhile production of nuts. Coastal areas are usually good sites whereas hill slopes or excessively wet sites are environments where coconut should not be grown. Coconut is found to a lesser extent in agriculturally-rich environments because other crops are more productive. Unlike many other crops, coconuts can grow well in areas of high rainfall, where the soil drainage is good, or equally well in areas of low rainfall, if there is a good supply of ground water. It tolerates a greater degree of soil salinity than many other plants. It provides food and drink, fuel and shelter, and the possibility of a cash income, often in harsh environments. The wide tolerance shown by coconut is due in part to its adaptation in the wild to particular environments before humans arrived (Harries, 1992a,b). Given the tolerance of coconut to a wide range of environmental stresses, it is not surprising that coconut was planted extensively both on plantations and on smallholdings.

Coconut fits an ecological niche not suitable for other plants. The coconut palm grows on coastal sands that do not suit oilpalm, and under rainfall that is too wet for soyabean. Much, or all, of the energy needed for processing could come from the production of shell charcoal, a valuable product itself. Economic production from a coconut enterprise can be enhanced by intercropping, underplanting, or grazing. Old trees need to be replaced by locally adapted, high-yielding varieties,tolerant to local pests and diseases, in order to raise the yield level.

Despite the long history of coconut cultivation, the standard of cultivation is generally low. Coconut is able to continue to produce some nuts even under conditions of poor management. Consequently, it is often referred to as a 'poor man's crop'. A low input/output system for coconut has come to be recognized as standard in many countries. Yet, coconut also responds to good management, and improved nutrient levels, and has an impressively high yield potential.

Coconut types

Coconut has two naturally occurring types, the tall and the dwarf. In addition, coconut hybrids, mainly resulting from tall × dwarf crosses, have been bred in many countries. Tall × tall hybrids are also possible, and have certain desirable characteristics, such as large nuts.

The common tall is a hardy type, which can live for at least 90 years. The economic life is generally considered to be about 60 years, depending on productivity and the price of coconut oil. Most of the world's coconut population is made up of the talls. The tall type has a single, unbranched trunk, growing to about 30 metres in height. The crown has 25–40 fronds, with a fully opened frond being about 6 m in length. Bearing begins after 5–7 years, reaching a peak at 15–20 years. Under favourable conditions, it will produce 60–70 nuts per year, and the nuts mature within 12 months of pollination. The mature nuts consist of approximately 35% husk, 12% shell, 28% meat and 25% water (Purseglove, 1975). The dwarf variety is characterized by its short stature and early bearing. It starts bearing after 3 years, and reaches full production within 6 years.

Hybrid varieties combine the characteristics of both types, with their chief economic advantages being their early bearing and high yield at maturity. Under favourable conditions, hybrids will produce up to 160 nuts per year (30 kg copra). They usually begin bearing after 4 years, and reach peak production within 20 years.

Coconut-based Agroforestry Systems

A large number of compatible crops, both annuals and perennials are grown under coconuts in different geographic and climatic regions. Coconut-based agroforestry systems were surveyed as part of an ICRAF-sponsored agroforestry systems inventory (Nair, 1983, 1989). Coconut intercropping systems have been studied in detail in India (Nair, 1983), and Sri Lanka (Liyanage *et al.*, 1989). Cattle grazing under coconut is important in Asia and the South Pacific (Plucknett, 1979; Smith and Whiteman, 1983).

There is a competitive relationship between species in mixed culture. During a time of plentiful water supply, light interception may be the limitation to growth. If coconut is planted at a density for maximum yield, the canopy will intercept a high proportion of light unless the density has been reduced to accommodate severe annual water deficit. In that case, there may be a good opportunity to grow a short-lived intercrop during the wet season. Where high density prevails, there is little transmitted light available for up to 20 years, after which there is a gradual rise in the transmitted light, to the benefit of the intercrops (Foale, personal communication).

The characteristics, constraints, and research needs for some illustrative coconut-based agroforestry systems in Asia, the South Pacific and the Americas are described in the following sections.

Asian systems

Cropping patterns

Intercropping of coconut with other food and cash crops is common throughout south and southeast Asia. In Sri Lanka, for example, bananas, black pepper, coffee and ginger are the intercrops most preferred by farmers. The second group (in order of preference) consists of turmeric, betel, vegetables and pineapple. Factors such as profitability, marketing facilities and convenience are the major reasons for farmers' preferences for these crops (Liyanage *et al.*, 1989).

Coconut stands can conveniently be intercropped either when they are young (up to 8 years after planting) or fully grown and showing a decline in light interception (from about 25 years after planting). Theoretical considerations of plant–community interactions in multi-species combinations with coconuts have been discussed by Nair (1989). The main expectation from an intercropping system in a perennial plantation-crop system is that the overall return from a unit of land is increased without adversely affecting either the current or the long-term productivity of the main (perennial) crop. At the same time, the returns from the additional crops should justify the adoption of the intercropping practice and should contribute to the long-term productivity of the system. Thus, intercropping in coconut stands is viewed as a means of increasing the total productivity of lands that are 'committed' to the coconut crop for at least 70 years (which is the normal life span of the tall type most commonly grown throughout the world) (Nair, 1989).

Results of intercropping experiments conducted at the Coconut Research Institute in Sri Lanka indicate that intercropping results in an increase of nut yields of coconut. Similar reports are also available from intercropping trials in India. The explanation given for this beneficial inter-action is that the palms benefit from the manure and other fertilizers given to the intercrops, elimination of weeds, soil working and other management practices (Liyanage *et al.*, 1989).

In Sri Lanka coconut is generally grown under rain-fed conditions. Experimental evidence shows that there would be no serious competition for soil moisture between the coconut palms and the intercrops if the annual rainfall exceeds 1900 mm.

Coconut-based intercropping systems have great scope for expansion to other areas. In principle, this is a sustainable system provided that necessary inputs are available at the appropriate times and in the correct quantities, and the system is managed efficiently.

Labour

Labour is one of the other major resources needed for intercropping. A study in Sri Lanka revealed that intercropping in coconut stands resulted in a 300% increase in on-farm employment. It shows that, depending on the type and number of intercrops involved, the requirement of labour and the share of labour cost in the total cost of production increased. While the timely availability of labour could pose a problem in some places, the generation of additional on-farm employment can be an encouraging aspect in owner-cultivated smallholdings (Liyanage *et al.*, 1989).

Soil conservation

Monocrop stands of coconuts offer only partial shading of the ground when the palms are young, and also as they advance in age and height and the crowns become smaller. Consequently, unless there is a secure coverage of herbage plants, the soil is more exposed to erosion and degradation during these periods. These are also the periods when intercropping is most feasible and desirable. In monocrop coconut stands, it is a common management practice to adopt soil and water conservation practices such as terracing, preparation of bunds and contour drains and burying coconut husks in pits and trenches near the palms to conserve moisture. By practising intercrop-ping and adopting prudent land-management practices for the intercrops, many of these soil conservation practices which would otherwise be necessary could be avoided. Thus, intercropping can be a better way of increasing the sustainability of coconut lands (Nair, 1989). The risk of reducing the physical quality of the soil surface due to the accelerated breakdown of organic matter that accompanies cultivation should be assessed carefully.

Constraints

The sample survey of the intercropping in coconut lands in Sri Lanka has identified several important constraints that are faced by the farmer when expanding his/her intercropping activity. These, in order of relative importance, are: (i) drought; (ii) lack of funds; (iii) price instability; (iv) lack of technical know-how; (v) problems of timely availability of labour; and (vi) insufficient planting material (Liyanage *et al.*, 1989).

In addition, marketing of perishable seasonal crops (e.g. passion fruit, papaya, pineapple) and crops that are produced in bulk (e.g. ginger, turmeric) can also be a serious problem as the prices are highly volatile. The problem can be aggravated if intercropping extends to a large area without simultaneously developing processing facilities at the producing centres and/or a transport system to consuming or processing centres.

Research needs

The constraints call for research on both biological and socioeconomic aspects, and the development of an efficient extension service in order to make coconut intercropping systems more productive, economical, adoptable, and successful. The agronomic requirements of individual crops when they are grown as intercrops need to be understood. The interaction of crops when they are grown in close proximity needs to be studied so that research can focus on the pattern of sharing growth-limiting resources by the component species of the system.

In order to arrive at prudent management recommendations it is necessary to take into account both complementary and competitive interactions affecting production of individual species as well as total production of the whole system over time. A reassessment of the presently accepted planting pattern and density of sole crop coconut is also worth undertaking with the objective of growing intercrops without adversely affecting the palm productivity. Ways of maintaining soil fertility in coconut intercropping systems through the application of nutrients need to be identified. Research on various aspects of the related system of pastures with animal production under coconuts also needs to be intensified (Liyanage *et al.*, 1989).

South Pacific systems

Some forms of agroforestry exist in almost all countries of the Pacific region. Although there are some broad similarities among the countries with respect to their agroforestry situations, there are also many differences among them due to the variations in the physical, socioeconomic and land-use conditions. The various systems are described by Schirmer (1984) and Vergara and Nair (1989). The main systems are summarized below.

Coconut, cacao and food-production systems

In the coastal plains throughout Papua New Guinea and on the Pacific Islands, the ubiquitous perennial crop is coconut, the basis of the copra trade for almost a century. With spacing varying from 8 to 10 m between palms, other useful species are grown between and under the palms. For example, intercropping cacao among coconut has become a promising practice. Certain sections of the farms often contain a third element, with root crops (such as sweet potato, cassava or taro) as a supply of staple food.

On farms where the introduction of cacao in existing coconut plantations arose from a late recognition of the potential of the wide and unutilized interspaces between the palms, the coconut palms were of fully fruit-bearing age (around 30 years) at the time of intercropping. However, new plantations involve the simultaneous planting of coconut and cacao (and sometimes root crops). Market oriented farms generally raise only coconut and cacao as cash crops, while smallholder farms are more oriented toward domestic use and include an appreciable food-cropping element. Special allowance must be made for this by planting the coconut palms wider apart, because the light interception of palms in the 8–20 year age range is much higher than in older palms (Wilson and Ludlow, 1991).

Coconut, legume tree, cacao production systems

A variant of the coconut–cacao–food-crop system is a three-tiered, all-perennial combination that excludes the annual food crop component, and substitutes a third perennial component, this being a legume tree species (usually *Leucaena leucocephala* in Papua New Guinea, and *Glificidia septum* in the Solomon Islands). The major functions of the legume trees are to: (i) provide partial shade to cacao plants; (ii) serve as a source of fuelwood; and (iii) to enrich the soil through symbiotic nitrogen fixation and the addition of leaf litter. This system is now widespread in Papua New Guinea, Solomon Islands and Western Samoa.

Silvopastoral: cattle grazing under perennial crops

Grazing cattle on pastures under coconuts to keep down the competing and fire hazardous grassy weeds has been found to be feasible, as demonstrated in the Solomon Islands, and the coastal areas and outer islands of Papua New Guinea, Tonga and Fiji. In this system, cattle contribute to the production of animal protein for local diets, recycle the biomass and also provide some nutrients in the form of cattle manure to improve soil fertility and increase farm income and employment. The Solomon Islands has a cattle-under-trees programme that combines coconut or plantation forest production with cattle (Smith and Whiteman, 1983).

Agroforestry as a means to reforestation

Due to excessive wood-extraction activities, and conversion of forest areas to unsustainable food cropping, considerable areas of hilly land have degenerated into unproductive grasslands, in Papua New Guinea, Fiji and the Solomon Islands. These lands could be returned to a productive and stable status through reforestation if governments could afford to cover the initial high costs. A relatively inexpensive alternative to reforestation is agroforestry, where farmers can establish tree crops and food crops simultaneously, and may be compensated in terms of the food output. They move to other open areas to repeat the tree-establishment process as soon as the tree crops reach canopy-closure age which precludes further food cropping (Nair, 1989).

Studies of human ecology have shown that the adoption of agroforestry practices is influenced not only by biological and environmental factors,but also by socioeconomic and cultural factors, particularly land tenure, population pressure and marketability of produce. The ratio of people to land resources is a convenient gauge of pressure on land, and the extremely limited land areas of the Pacific islands are now under human population pressure (Vagara and Nair, 1989).

Land tenure

The land-tenure pattern varies from country to country in the Pacific region. A pattern that is common to most is the traditional clan exercising extended-family control or ownership of land as opposed to the private individual or government ownership as in most southeast Asian nations. Clan-owned land may either be cultivated communally, with each clan member receiving a proportionate share of the output, or apportioned among the individual households of the clan and used in a semi-private manner. Under the latter arrangement, absolute ownership of the land continues to be with the clan. The houseowner merely acquires temporary rights to and control over that piece of land but holds absolute ownership over the cultivated plants. If the crops are perennial, the farmer may retain control over the land for an extended period and the arrangement may appear like a private form of ownership.

The influence of the clan-type of customary or traditional tenure system upon agroforestry could be either negative or positive. Where clans are not affected by heavy population pressure on land, and do not specify whether the crops cultivated by temporary assignees of clan land should be annual or perennial, the households could readily plant perennial crops in combination with their annual food crops. This is practised in the eastern highlands of Papua New Guinea. In addition to the perceived ecological and economic benefits of including trees on their farmlands, the farmers enjoy the added advantage of being able to prolong indefinitely their 'tenancy' over that piece of clan land for the duration of the perennial crops. In this case, the clan-type land tenure favours and encourages agroforestry practice.

The opposite situation could arise when clans prohibit the raising of 'permanent' perennials, such as fruit trees, for fear that clan lands will be fractured into small plots and taken from clan control for permanent use by individual households. This fear can be overcome in the form of a clan-controlled rather than an individual household-owned agroforestry farm. The clan's fear of having the lands gradually transferred to private control would be eliminated (Vagara and Nair, 1989).

Marketability of produce

Achieving the twin goals of productivity and sustainability through agroforestry could result in the production of goods in excess of local demands, which need to be marketed domestically or abroad for cash incomes. Local sale of sweet potato, papaya, taro and cassava, and export sales of coconut, cacao and coffee from many Pacific countries are examples of enhanced economic activities. Accompanying problems may also arise which could dampen the enthusiasm of agroforestry practitioners. For example, lack of ready access to markets can be a problem. The export-oriented outputs of agroforestry may not be of significant volume in many of these small countries to attract international carriers, thereby cutting off their access to international markets. Thus, inability to dispose of the surplus outputs profitably is a disincentive to agroforestry in this region (Vagara and Nair, 1989).

Improving physical access to markets requires time and capital resources which many island countries do not possess. Fortunately, the types of products from these countries provide a great degree of marketing flexibility which could outweigh the problems of poor access. For example, with minimal processing, the storability of exportable products like coffee and coconut can be extended so that when a sufficiently large volume accumulates international shippers could make port calls and collect the goods. This is now happening for coffee, cocoa and copra in Papua New Guinea.

American systems

The common feature which links the three tree crops, cashew, coconut and carnauba in northeast Brazil is livestock grazing (silvopastoral system). Cultivation of annual subsistence crops, and other perennials, in the stands of these three perennial crops is also common (Johnson and Nair, 1989).

Both cashew and carnauba are native to northeast Brazil and wild stands have a long history of exploitation. Coconut is a more recent introduction. In addition to gathering the economic products of these three trees, the land areas they occupy are also traditionally used for other agricultural purposes. Locally raised livestock (cattle, goats, and donkeys) are grazed on spontaneous grass and shrub growth beneath the trees and so are cattle brought in from the interior for seasonal grazing. These tree stands provide much-needed shade for the livestock. On small farms, coconut and cashew, or cashew, banana and coconut are interplanted.

The perennial crop-based system has considerable merit in the context of the environmental, agricultural and socioeconomic conditions in northeast Brazil. From an environmental standpoint, establishment of new plantations of cashew or coconut represents an upgrading of the vegetative cover from what currently exists. Plantation grazing adheres to the general recommendation that grazing is a sustainable system for the dry savannah climate.

The diversity of the perennial-crop-based agroforestry system offers the advantage of making more efficient use of labour and equipment over the entire calendar year, thereby avoiding the peaks and slacks of activity associated with monoculture. Plantation grazing does, however, have a few disadvantages. Careful management practices must be followed to prevent animals from physically damaging young trees.

There are a number of socioeconomic benefits from this system. Coconut products enjoy high demand within Brazil, which provides an incentive for expansion of the crop. Economically, livestock raising is profitable in the northeast, and combining it with cashew or coconut should strengthen the economic base of the individual plantation. In social terms, the growth of agroindustry in the region creates new industrial employment. Given the large rural labour pool, there is not a strong incentive to mechanize plantation operations. In the rural areas, local inhabitants also benefit from having free or inexpensive sources of thatch and wood from the plantations. Social reasons may be the overriding justification for encouraging the expansion of perennial-crop-based systems (Johnson and Nair, 1989).

Research needs

Although the coconut-based system is practised over extensive areas in Brazil, and it has a number of merits, there is an almost total lack of information on its various management details. Practically no research, nor systematic survey or data collection, has been done on coconut-based systems in the Americas. No quantitative information is available on many of the basic aspects, such as the relative use of light and nutrients by the components of the system. Other essential information includes quantitative information on the functional and dynamic aspects of production of the various components, as well as their rate of change of production with time. The next step would be to examine the reasons for the observed behaviour and see how the efficiency of production could be improved. This will involve research on a large number of management aspects of the individual components, as well as of the system itself (Johnson and Nair, 1989).

Box 2A: Edible and Non-edible Uses of Coconut

Copra

The main industrial use of coconut is in the production of copra, from which coconut oil and copra meal are derived. The initial step in copra making is the harvesting of mature nuts. The nuts are dehusked and split into halves and dried. Drying is done either through kiln or sun-drying. It is necessary to reduce the moisture content of the meat from about 50% to 5%, in order to reduce the weight, prevent microbiological deterioration, and allow concentration of the oil. On average, five nuts are required to produce 1 kg copra, but this conversion rate varies (plus or minus 40%) from country to country depending on the size of the nuts.

Coconut oil

Copra is further processed to obtain coconut oil and copra meal. Copra contains approximately 63% coconut oil, the remainder being meal, moisture and waste. Coconut oil consists of approximately 48% lauric acid, with the balance being mainly myristic, caprylic and palmitic acids (Ohler, 1984). The chemical composition of coconut oil allows it to be used for a wide range of edible and non-edible purposes. Its chemical composition is significantly different from other vegetable oils (apart from palm kernel), because of the high percentage of lauric acid, and other shorter chain fatty acids. This allows it to retain a market niche, despite competition from other oils such as soyabean and rapeseed oil.

Like most oils and fats, coconut oil serves as an important source of energy in the diet; it supplies specific nutritional requirements; provides lubricating action in dressings or leavening effect in baked items; functions as an effective heat transfer agent in frying; acts as carrier and protective agent for fat-soluble vitamins; and contributes to palatability and enhances the flavour of food.

Oil fats are the most concentrated of all food, providing about 9 kilocalories of energy per gram, against 4 or 5 kilocalories furnished by proteins or carbohydrates. Fat is the major energy source in developing countries, but its supply is limited in many countries. Coconut oil is a major source of energy in countries where the crop is grown. In the Philippines, for example, almost all cooking oil is 100% coconut oil.

Coconut oil is used for a wide range of edible and non-edible applications. For edible purposes, coconut oil has a bland flavour, pleasant odour, high resistance to rancidity, narrow temperature range of melting, easy digestibility and absorbability, high gloss for spray oil use, and superior foam retention capability for whip-topping use.

The most significant physical property of coconut oil, which sets it apart from most fats, is that it does not exhibit gradual softening with increasing temperature but passes rather abruptly from butter solid to liquid within a narrow temperature range. In this respect, it resembles cocoa butter. This melting behaviour is a consequence of the fatty acid composition of the triglyceride. Coconut oil is liquid at a temperature of 27°C or higher and solid at 22°C or lower. Since the quantity of unsaturated fatty acids is low, coconut oil

Box 2A: Continued

is extremely resistant to rancidity. Hydrogenation produces only a slight change in consistency and melting point of coconut oil.

Edible products from coconut oil

Cooking/frying oil   Coconut oil has a bland flavour and allows food cooked in it to retain its natural flavour. It is also extremely resistant to the development of rancidity, because of its low content (10%) of unsaturated oil, and has a long shelf life.

Shortening and baking fats   Although coconut has a high melting point (27°C), when combined with a small amount of hydrogenated palm or cotton-seed oil, it produces shortening of excellent quality. This shortening is used in the preparation of baked products, such as bread, cakes and biscuits.

Vanaspati and margarine   Coconut oil is used, in combination with other oils, to make vanaspati (a butter-like product used in South Asia) and margarine. Because coconut oil has the lowest level (9%) of unsaturated fatty acids of all edible vegetable oils, the cost of hydrogenation is much less than other oils. This low cost, plus its bland flavour and long shelf life make it ideal for use in many areas.

Substitutes for dairy products   Coconut oil is used as a substitute for more expensive milk fat in filled milk formulas and filled cheese. It is also used in the production of vegetable oil-based ice cream, coffee whiteners and whipped dessert toppings.

Spray oil   Coconut oil is used as spray oil for crackers, cookies, and cereals to enhance flavour, increase shelf life and give a glossy appearance.

Confectionery   Coconut oil has melting characteristics similar to cocoa butter, and is used extensively in confectionery.

Other edible applications   Coconut oil is used as a fat source in baby food because of its bland flavour and easy digestibility. It may also be used as a salad oil, if stored at higher than 27°C.

Non-edible products from coconut oil

Coconut oil is one of the few vegetable oils which can be utilized in a wide range of non-edible and industrial applications. It is often preferred to synthetic or petroleum oil-based materials because of its unique characteristics.

    For non-edible purposes, the desirable properties of coconut oil derivatives are due to: (i) the excellent resistance of the oil to randicity, leading to long storage life of derivatives; (ii) it produces detergent materials having a desirable combination of lathering solubility and detergency characteristics; (iii) its biodegradability (natural products are increasingly favoured by consumers over synthetic ones); (iv) its excellent qualities as a carrier of many types of active materials; and (v) its relatively non-oily character and mildness to the skin.

    Some of the important non-edible uses of coconut oil are described below. Of these items, the major non-edible uses are in soaps, shampoos, shaving cream and other cosmetics.

Box 2A: Continued

Soap, shampoos, and cosmetics   Coconut oil is used for making both laundry and toilet soaps. The high lauric acid content of coconut oil gives the soap good lathering qualities.

Chemical products   Coco-chemicals derived from coconut oil are used in a wide range of consumer products and industrial applications.

Fuel blender and diesel-oil substitute   Coconut oil can be used as a substitute for diesel oils, for electric generating plants and motor vehicles. However, this use is non-economic in most situations at the present prices of fuel oil.

Copra meal

Copra meal or cake is the residue remaining after the oil has been removed from copra. It contains approximately 20% protein, 45% carbohydrate, 11% fibre, and the balance is fat, minerals and moisture. It is used as a component in animal feed. Most is exported to Europe as pellets for use in cattle feed rations.

Desiccated coconut

Desiccated coconut is the dried, shredded coconut meat prepared for food use. The shell of the husked nut is removed, the thin brown skin of the coconut kernel pared off, and the coconut water removed from the nut. The white meat of the coconut is cut into small particles by machine, pasteurized by steam heating, dried, screened, graded and packed. Strict microbial quality control is necessary to ensure that a hygienic product is marketed, to meet export standards.

Desiccated coconut is used in the preparation of edible products in five areas:

1. the confectionery industry, as the main flavour ingredient in chocolate bars or as a filler in nut-based products;

2. the bakery industry, as a flavour ingredient or as a filler in cakes and biscuits;

3. frozen food, as a decoration for ice cream and other frozen food products;

4. the general food processing industry, as a flavour and filler in packaged and canned foods;

5. the consumer product industry, as an ingredient in ready-to-cook mixes, and packaged desiccated coconut for home use.

Other edible coconut products

Several other edible products are made from coconuts. They are not yet major items of world trade, but they are important for consumption in the producing countries. There may be prospects for marketing them more widely, these are as follows.

Fresh coconuts   Fresh coconuts provide coconut meat for use in a wide variety of dishes, either as a snack, or in the preparation of main courses and desserts.

Box 2A: Continued

Young fresh coconuts contain coconut water as a refreshing drink. The water can also be mixed with other beverages. Fresh coconuts are sold widely in producing countries. They are exported in small volumes to the United States, Europe, Australia and New Zealand, amongst others.

Coconut chips   Coconut chips are thinly sliced, dried, mature coconut meat. They can be used as a snack food, as a substitute for potato chips and similar products.

Coconut milk and coconut cream   Coconut milk (coconut cream) is the filtered white extract from grated coconut meat. It is a common ingredient in many Asian and Pacific recipes, particularly curries and fish dishes. It is now available in tins, tetra-paks, and in powdered form. There is the potential to expand the sales of these products worldwide.

Coconut flour   Coconut flour is a clear white, food-grade residue from pressed and/or solvent extracted, dehydrated coconut meat. The average protein content is 24%. Its inclusion with wheat flour increases the food's protein value.

Coconut jam   Coconut jam is made by mixing coconut milk with brown sugar. It is a popular food spread in the Philippines.

Other food products   Coconut honey, coconut candies and several coconut desserts can be made from coconut meat. These products are especially popular as snacks in the Philippines. In Indonesia, a 'tempeh' product has been developed by the fermentation of coconut milk using a suitable bacterial culture.

Coconut-shell charcoal and activated carbon

Coconut shells have a similar composition to hardwoods, but have a higher lignin and lower cellulose content. The coconut shell is used locally as fuel for copra making and for the kitchen fire in producing countries.

The main commercial product is charcoal. Coconut shell charcoal processing involves the burning of shells of fully matured nuts, in a limited supply of air. Coconut shell charcoal is the major source of domestic fuel in the Philippines. It is also exported by the Philippines, mainly to Japan, USA and Europe.

Coconut shell charcoal may be processed further into the high value product, activated carbon. Activated carbon acts as a high grade filter. It has many industrial applications, including: general water purification to remove free chlorine, organic matter, iron content, taste and odour; crystalline sugar preparation; contact catalysts and carriers; solvent recovery in dry cleaning; decolonizing and deodorizing agents; gold purification.

Coir fibre

Coir fibre is extracted from the husk of coconut. White fibre is obtained by retting fresh green nuts, preferably in saline water. The fibre is spun into yarn, and used in the manufacture of mats, carpets, and rope. Brown fibre is obtained

Box 2A: Continued

mechanically from brown husks by milling. This produces long, stiff bristles for use in brushes, or medium to short fibres for use as mattress and upholstery filling.

Coconut wood products

Coconut timber taken from the lower and middle parts of the trunk can be used for load-bearing structures in buildings, such as frames, floors and trusses. Coconut trunks can be used for poles, as they have great strength and flexibility. Coconut wood can also be used for furniture and parquet flooring.

Box 2B: Harvard Medical School Testimony to the US Senate on Tropical Oils

Mr Chairman and Members of the Committee:

Thank you for this opportunity to appear before you today and to alert you to the possibility that our quest for a healthy American diet is being distorted by an unwarranted attack against tropical oils.

My testimony will focus on coconut oil, also known as the lauric acid group. Because this particlular tropical oil (along with palm kernel oil) is saturated, assumptions have been made that it has an adverse effect on blood cholesterol. In fact, the fairest interpretation of the scientific literature is a neutral effect, even in situations where coconut oil is the sole source of fat. In most cases, coconut oil makes up less than 29% of the calories per day, thus making a trivial contribution to fat intake.

Coconut oil can also be seen as a specialty oil that plays an important role in our food products. It increases shelf-life in most products and thus reduces the cost of production, and it adds to the flavor (principally, in biscuits, cookies and other baked goods) and thus satisfies the consumer.

The reason for these two phenomena is that this particular vegetable oil contains shorter chain fatty acids which affect the oil's melting point. Therefore, as a cookie changes from room temperature to body temperature, a cooling effect occurs that enhances the taste. At the same time, the body requires no transport system to absorb, digest or metabolize this oil. Therefore, virtually none of it is incorporated into body fat, and none of it is available for the synthesis of cholesterol.

It is also important to note that coconut oil has long been an important component of medical foods and baby foods. Research in our laboratory is demonstrating that coconut oil may be the preferred fuel for individuals sustaining serious illness, including burns, sepsis, malnutrition, and immunologic problems, including AIDS. We are currently investigating the effects of lipids on endotoxemia, which is particularly prevalent in patients with AIDS and other immunologic deficiency syndromes. We have shown the superiority of coconut oil in comparison to polyunsaturated oil such as corn oil.

It is now apparent that selective labelling requirements that focus on the lauric acid tropical oils (coconut and palm kernel) would represent an inaccurate and misleading use of existing data. **There is simply no scientific basis for describing coconut oil as a health risk**.

Mr Chairman, those who would claim that coconut oil or palm kernel oil pose a threat to the consumer are relying on misinformation allegedly derived from statements contained in the US Dietary Guidelines and recommendations from health organiazations such as the American Heart Association. The focus of these recommendations, however, is on weight control through calorie control, the reduction of total fat intake to less than 30% of calories and control of cholesterol. Specifically, fat intake is recommended to be ⅓ saturated, ⅓ monounsaturated, and ⅓ polyunsaturated. A focus on the so-called 'omega-6' vegetable oils versus the tropical oils would yield trivial results and certainly would not be supported by scientific data dealing with the US diet where these oils play such a small part in the dietary intake.

Box 2B: Continued

Any valid conclusion concerning the health effects of oils must take into consideration the chemical structure for the saturated oils and must recognize that the medium chain lauric acids (C6 to C12) are quite different from the long chain (C16 to C18). Coconut and palm kernel oils derive more than ⅔ of their fatty acids from lauric acids or shorter chain fatty acids. For more than 30 years, it has been known that the group of 'medium chain triglycerides' have properties very different from long chain triglycerides which predominate in soybean and vegetable oils. Saturated fats composed principally of medium chain triglycerides (MCTs) do not elevate serum cholesterol when taken as part of the normal diet. The chemical properties of MCTs are such that they are rapidly metabolized, and therefore contribute a preferred energy source for the body. As a result, unlike other saturated fats, they are not stored in the body as fat.

Soybean oil, by contrast, is a long chain triglyceride. In its natural state, it is polyunsaturated of the omega-6 category. When soybean is hydrogenated, however, it becomes more like animal fats, which are long chain triglycerides of the saturated variety known to contribute to blood cholesterol, although in a modest to trivial fashion. Typically, soybean oil is hydrogenated when used as a substitute for coconut oil and does tend to elevate cholesterol levels. Coconut oil, in contrast, whether in its natural state or hydrogenated, still has the majority of its fatty acids as medium chain triglycerides.

As already noted, ideal dietary fat intake combines a mixture of three main fat groups, as reflected in the US Dietary Guidelines and the American Heart Association's recommendations. In practice, since coconut oil is only a tiny fraction—less than 1%—of the US dietary fat intake, and since it is a mixture of mono and polyunsaturated fats, it is a very unique and desirable fat for human use.

It is interesting to consider how misinformation regarding lauric acid oils originated. Largely, this arose from the fact that coconut oil has always been cheap and easy to use in animal experiments investigating the basic science process of atherosclerosis. While such experiments were helpful in these investigations, they bear no relationship to the topic of a healthy heart diet. The principal reason is that in atherosclerosis investigations, 2–3% fat calories must come from linoleic acid to satisfy the essential fatty acid requirement. Yet in the experiments that have been used as the foundation for labeling coconut oil as a 'bad fat', this linoleic acid component was usually missing. When you study coconut oil as part of a diet that does contain linoleic acid, as the American diet does, you discover that coconut oil does not affect the process of atherosclerosis any differently than any other vegetable oil, even though it is high in saturates. Thus, for the US consumer, the use of coconut oil does not increase the risk of heart disease.

In contrast, hydrogenated soybean and other vegetable oils do tend to raise cholesterol levels. Moreover, if those vegetable oils are consumed in a natural state as part of a diet that derives more than 10% of its calories from polyunsaturated, they may increase the risk of various types of cancer. We have

Box 2B: Continued

provided a detailed review of the animal and human studies to document these facts.

When one combines our studies with the epidemiologic data developed by Dr Quintin Kintanar, Executive Director of the Philippines Council on Health Research and Development, the findings reveal that no relationship exists between coconut oil consumption and heart disease mortality. Dr Kintanar's findings show that in the Philippines, where coconut oil is a far greater portion of the daily dietary fat intake than in the US, the rate of cardiovascular disease is much lower than in the US.

In the Philippines, a diet deriving 6% of calories from coconut oil correlates to an incidence of heart attack mortality of 22 per 100,000. In the US, the intake of coconut oil is less than 1% of dietary calories; yet the incidence of heart attack mortality is 227 per 100,000 – ten times the incidence in the Philippines. Further neither in the Philippines nor the US is there a proven relationship, either statistically or as a matter of cause and effect, between coconut consumption and heart disease. Indeed the average coconut oil consumption in the US is less than 2% of the total annual edible oil consumption of 16 million tons. While the trend in US mortality due to cardiovascular disease is declining, the consumption of coconut oil remains stable, further challenging the focus on coconut oil as a health risk.

In closing, I wish to emphasize that a great deal of research remains to be done in analyzing the effects of dietary fatty acids on human health. It is not a subject that is yet ripe for legislation of the sort under consideration today. It would be particularly unfortunate if consumers were deterred from buying products containing coconut oil on health grounds, when the most recent medical evidence suggests that coconut oil is more beneficial to consumers than the hydrogenated fats that would be exempt from the proposed legislation. Thank you.

Source: *Coconuts Today* 30 December 1987, pp. 22–23

Testimony before the US Senate Committee on Labor and Human Resources on S.1109, a bill to amend the Federal Food, Drug and Cosmetic Act to require new labels on foods containing coconut, palm and palm kernel oil by : George L. Blackburn, Vigen K. Babayan, Beatrice Kanders, Nawfal W. Istfan, Marilyn Kowalchuk, Harvard Medical School, Nutrition Coordinating Centre, New England Deaconess Hospital, 194 Pilgrim Road, Boston, MA 02215. 1 December 1987.

Chapter three:
World Fats and Oils Market

'The future prospects for coconut oil are quite favourable,
because of the growing demand for vegetable oils . . .'

Overview

The worldwide prevalence of oilseeds, the combination of annual and perennial crops, the jointly produced products, and the wide-ranging degree of
technical substitution among oilseed products produces a unique and intri-
cate market with a number of distinctive economic and physical characteristics. As described by the World Bank International Economics Department
(1991), worldwide demand for both oils and meals will grow most rapidly in
the developing countries. Driven by population and income growth, the
demand for both oils and meals is growing rapidly and will continue to do
so through the 1990s. The demand for both sets of products has grown most
rapidly in the countries where income gains have led to better and more
diversified diets. As incomes climb in the poorest of countries, consumers
quickly increase their consumption of cooking oil to add flavour, calories,
and diversity to their diets. When incomes climb still further, poultry and
other meat sources are added to the diet, contributing to the demand for
oilseed meals.

The consumption of oilseed products has grown most rapidly outside of
the major producing areas, thus stimulating international trade in these
products. Oilseeds and their products are traded in a worldwide market, the
workings of which are complicated by the physical nature of the production
processes for both raw and finished products. Production includes both
annual crops grown on land easily used for alternative crops and perennial
tree crops. Some production technologies are quite input-intensive, while
other growing methods—especially for some tree crops—require no
additional inputs other than land, rain, and harvesting labour. All countries

consume oilseed products, in forms ranging from animal feed, to cooking oil, to industrial products such as detergents and paint products. Oilseeds contain multiple products, and all oilseeds contain both oils and meals in varying amounts. All meals can be used as feed, but each possesses a unqiue nutritional composition characterized especially by protein content. Almost all of the vegetable oils can be used as cooking oil, and, through various methods of processing, all oils can be broken down into their basic chemical components and recombined to form nearly homogeneous final products. Because of the multiple and varied nature of final oil products, vegetable oils compete in markets shared with animal products such as butter and tallow, and in markets shared with petroleum products such as detergents and wood finishes (World Bank, 1991).

Structural changes occurring in the oilseed markets during the past decade will have lasting consequences. The emergence of the EC as a significant producer of oilseeds and protein meals, the rapid increase of productive capacity for palm oil in Malaysia and Indonesia, and expanded soyabean plantings in South America have brought an increased competitiveness to world markets that will be reflected in a downtrend in real prices over the long term.

With shifting populations, low prices, and a significant reduction in the incidence of poverty forecast for Asia, the prospects are good for substantial gains the demand for oilseed products. Growing populations and strong gains in income should boost total and per capita demand for both vegetable oils and meals throughout all of Asia. Consumption is expected to grow fastest in countries where demand levels are most sensitive to income gains and population growth rates are highest. Steady growth is projected for the mature markets of Europe, North America, and Japan. Significant gains in vegetable oil demand are also anticipated in Eastern Europe, where vegetable oils are expected to displace animal fats as internal prices in those countries move to reflect international levels.

Worldwide production of oilseeds is diverse, and the diversity will continue to increase. Supplies of low-cost soyabeans from South America and palm and palm kernel oil from southeast Asia are expected to continue their rapid expansion. Production in the EC, which grew quickly through the early 1980s is expected to slow; United States production levels are not expected to change dramatically. Export markets for African groundnut oil and meal are expected to stagnate. Coconut supplies are expected to remain steady or shrink in most coconut-producing countries, including India, Malaysia, Sri Lanka, and the Philippines. Production and exports are projected to grow only in Indonesia due to smallholder production gains among the Outer Islands (World Bank, 1991).

A more detailed discussion of demand and supply forecasts, world trade in the fats and oil market, price predictions, and future prospects for coconut and palm kernel oil is given in the following sections.

Demand Forecasts

Edible oil is a major component of the diet, and an important earner of foreign exchange for many countries. Demand for fats and oils is increasing faster than world population growth. In an analysis of the world market for oils, fats and meals from 1958 to the year 2000, Mielke (1988) noted that between 1982–1987, world population increased by 8.7%, while per capita demand for 17 fats and oils increased by 10.3%.

Demand for oilseeds and their products is influenced by a variety of factors, including population and income growth, relative prices, cultural tastes and preferences, and government policy. There is a pattern of demand development which transcends national boundaries. At a low level of income, when per capita calorie supplies are around 2000 calories per day, vegetable oils are often considered a luxury. This was true in Bangladesh, China, and the Philippines in the early 1980s, when calories from vegetable oil sources composed only 2–3% of the diet. As per capita income increases, the daily demand for vegetable oils in caloric terms grows rapidly from levels below 100 calories per capita to 200 calories per day. At still-higher income and a per capita calorie level in excess of 3000 per day, the demand for vegetable oils tends to taper off. The absolute level of consumption may well depend on taste and cultural preference, e.g., it is relatively high in Italy and the United States and more moderate in France and Switzerland (World Bank, 1988).

These trends are reflected in Fig. 3.1, which shows the per capita disappearance of fats and oils between 1958 and the year 2000, for the major consuming countries. In 1958, the world average disappearance of oils and fats was 10 kg. In 1983 it was approximately 13 kg. By 2000, it is expected to be approximately 18 kg. Much of this growth will come from an above-average increase in per capita consumption in countries such as Brazil, China, India and Indonesia (Mielke, 1988).

The actual consumption in 1980, and the projected demand in 1990 for fats and oils in various geographic regions are given in Table 3.1. It is estimated that total consumption will grow by 4.2% p.a. in developing countries, compared with 1.6% p.a. growth in industrial countries and 2.4% p.a. in the centrally planned economies (IRHO, 1986).

The per capita disappearance of oils and fats between 1958 and 2000 by region is illustrated in Fig. 3.2. The major consumers are: (i) the European Community countries; (ii) other industrial countries; (iii) China and India (Fig. 3.1).

The growing deficit in oils and fats between 1987 and 2000 is illustrated in Fig. 3.4. The areas expected to have a major deficit by the year 2000 include China and India. The countries which are projected to have excess production available for export are USA, Canada, Argentina, Brazil, Malaysia and

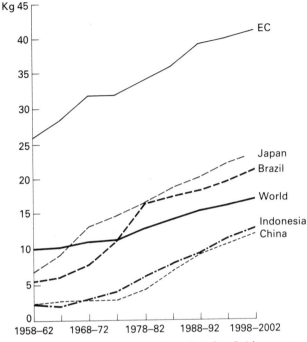

Fig. 3.1. Per capita disappearance of 17 oils & fats (kg) by country (source: Mielke, 1988).

Indonesia (Fig. 3.5). Based on current assumptions, it is estimated that total world consumption of fats and oils will be 104 million tonnes by 2000 (Fig. 3.3). This would be an increase of 42% on 1987 consumption. It means that on average an additional 2.4 million tonnes will be needed every year (Mielke, 1988).

Demand for fats and oils is increasing primarily in developing countries. It is expected that developing countries will account for 50% of total consumption by the 1990s, compared to 45% in 1980, and 36% in 1972 (IRHO, 1986). The major increases in demand are in China, Indonesia and Brazil. In the industrial countries, there is little prospect for a significant increase in the consumption of fats and oils, since the income elasticity of demand is low, the population is not increasing, and there are some perceptions of health risk in fats and oils (particularly from saturated fat as a contributory factor in heart disease). The net effect of these projections is that an increasing proportion of world production of fats and oils will be marketed in developing countries, where consumption is low, and demand is rising.

Growth in demand for vegetable fats and oils has been quite slow in the EC and the United States over the last few years. There is still room for substantial change among the various sources of fats, especially between animal and vegetable sources. Among the industrial countries, growth

Table 3.1. Fats and oils, actual consumption and projected demand, by region, 1980–90.

	Total consumption		Per capita food		Growth rates Total consumption	
	Actual 1980	Projected 1990	Actual 1980	Projected 1990	1972–80	1980–90
	('000 tonnes)		(kg per year)		(% per annum)	
World	58878	77602	10.8	12.0	3.4	2.8
Developing	25577	38646	6.5	8.2	5.9	4.2
Africa	3913	5652	8.3	9.2	5.5	4.0
Latin America	5517	8067	11.2	13.5	5.9	3.9
Near East	3239	4869	13.1	15.1	7.9	4.2
Far East	8663	13592	6.2	8.0	5.6	4.6
Asia CPE	4295	6417	3.3	4.5	5.4	4.1
Other Developing	50	50	7.0	5.5	7.1	− 0.1
Industrial	33301	38956	22.7	24.3	1.9	1.6
North America	8789	10176	29.5	30.5	1.9	1.5
Western Europe	12578	13664	25.9	26.6	1.6	0.8
E. Europe and CIS	8667	10982	18.1	20.8	2.1	2.4
Oceania	538	629	22.0	23.3	5.3	1.6
Other Industrial	2728	3505	15.1	17.1	2.0	2.5

Source: IRHO (1986)

prospects in Japan are promising. Throughout the past two decades per capita apparent consumption in Japan has grown, representing both income advances and an evolving change in preferences (World Bank, 1988).

Demand is expected to increase in Eastern Europe and the Commonwealth of Independent States (the former USSR). However, during the present transition phase in their economies, there has been a decline in oilseed production and consumption in 1990–91. Oilseed imports are expected to increase once foreign exchange credits are available (Oilworld, 1992).

Potentially, the largest sources of growth in demand are China and India. Each country possesses a large population. Income and caloric intake have improved greatly in recent years, yet both countries consume a below-world-average level of fats and oils. Imports of vegetable oils are controlled. Retail prices in India remain substantially above world market levels. Shortages have been reported in China in recent years, where apparent consumption of vegetable oils has grown rapidly since 1970. It is likely that even at current income levels, potential demand within both countries is stronger than the apparent consumption rates would indicate. Additionally, World Bank income growth forecasts are high for both countries. The implications are substantial. India and China represent more than one-third of the world's population. A 1 kg per capita increase in apparent consumption in the two

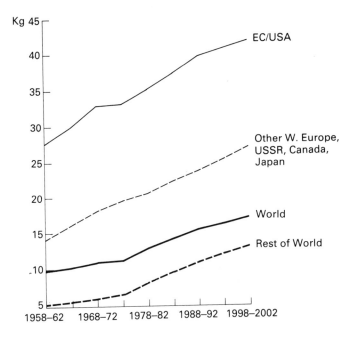

Fig. 3.2. Per capita disappearance of 17 oils & fats (kg) by region (source: Mielke, 1988).

countries would translate into more than 1.9 million tonnes of vegetable oil. This is more than double Indonesia's exports of palm oil for 1988. Increases of 2–3 kg per capita are certainly possible for both countries if incomes grow at a steady pace. This would bring additional pressure to bear on the governments of China and India to relax current high-price rationing policies (World Bank, 1988, 1991).

Brazil's total consumption of vegetable oils grew rapidly throughout the 1970s. At a per capita consumption of 18.7 kg per year, given projected population and income growth levels, Brazil will continue to be a source of strong demand well into the 1990s. Indonesia also represents another source of growth in demand. Despite substantial gains to date, the per capita consumption level remains low. Rising incomes and population growth should result in vigorous growth in demand for the next 10 years (World Bank, 1988, 1991).

As total world demand grows more rapidly in developing rather than industrial countries, there is likely to be a continuation of several trends in the composition of total world demand for fats and oils. Over the past two decades there has been a steady loss of market share for animal-source fats and oils (dairy products, lard, tallow, etc.) in industrial countries. More

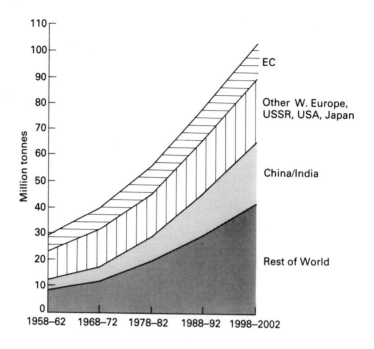

Fig. 3.3. The world disappearance of 17 oils & fats (source: Mielke, 1988).

significantly, the changing composition of demand between animal and vegetable fats and oils reflects a shift in final demand to countries that do not have a tradition of consuming animal-source fats and oils. As demand growth in Western Europe, North America, and parts of Latin America slows, and demand in Africa and Asia grows, there is likely to be a continued decline in the share of the market devoted to animal fats. In addition, the share of the world consumption devoted to palm oil should rise. Soyabean oil's share of the market is likely to remain fairly stable.

The options for developing countries to meet the increasing demand for fats and oils are either to import them, (using foreign exchange earned from other exports) or to increase local production. The long term projections on population growth and income growth suggest that there is a need for substantial production increases in developing countries in order to provide an adequate amount of fats and oils in the diet.

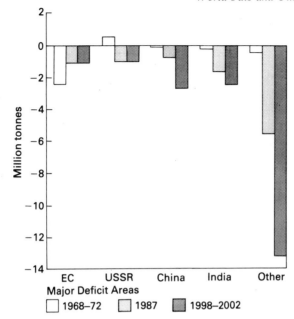

Fig. 3.4. The major deficit areas of the 17 oils & fats (source: Mielke, 1988).

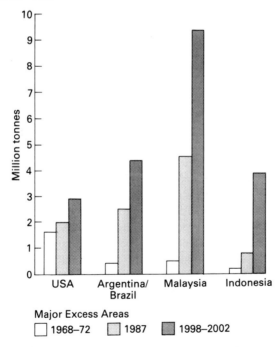

Fig. 3.5. The major excess areas of the 17 oils & fats (source: Meilke, 1988).

Table 3.2. Fats and oils, actual and projected production, by region, 1972–90.

	Actual			Projected	Growth rates	
	1972	1980	1983	1990	1972–80	1980–90
		('000 tonnes)			(% per annum)	
World	48876	60001	66178	77661	3.7	2.6
Developing	18010	25462	29033	39385	4.4	4.5
Africa	3093	3020	3119	3773	− 0.3	2.3
Latin America	3757	6594	6618	9388	7.3	3.6
Near East	1226	1504	1595	1836	2.6	2.0
Far East	7104	10275	11096	18098	4.7	5.8
Asian CPE	2642	3790	5473	5881	4.6	4.5
Other Developing	188	279	322	411	5.1	3.9
Industrial	26865	34539	37145	38276	3.2	1.0
North America	11896	17435	18579	18427	4.9	0.6
Western Europe	5876	7344	8401	8534	2.8	1.5
E. Europe and CIS	7489	7805	8295	9092	0.5	1.5
Oceania	792	946	979	1084	2.2	1.4
Other Industrial	813	1009	991	1140	2.7	1.2

Sources: FAO Production Yearbooks; IRHO (1986)

Supply Forecasts

The data given in Tables 3.2–3.4 summarize FAO production data for 17 major fats and oils (soyabean, cotton, groundnut, sunflower, rapeseed, sesame, maize, olive, coconut, palm kernel, palm oil, butter, lard, linseed and castor oils, fish oils and tallow). The production data for 17 major fats and oils in the major economic regions between 1972 and 1990, and the growth rates in 1972–80, and 1980–90 for various regions are given in Table 3.2. The production of individual oilseeds (including coconut) between 1972 and 1990 is shown in Table 3.3. Oilseed production is given in Table 3.4.

FAO predicted that world production of fats and oils would grow by an average of 2.6% p.a. in 1980–90, compared with 3.7% p.a. in 1972–80. Production in the industrial countries between 1980 and 1990 was expected to grow at only 1%, compared with 4.5% growth in the developing countries in the same period.

Much of the future production increases will come from developing countries, particularly from Asia. FAO estimates that in the 1990s, approximately 50% of total world production will come from developing countries, in contrast to 40% in the 1970s. The major production increases are expected to come from China, India, Indonesia, and Malaysia. Malaysia

Table 3.3. Fats and oils, actual and projected production, by commodity, 1972–90.

	Actual			Projected	Growth rates	
	1972	1980	1983	1990	1972–80	1980–90
		('000 tonnes)				(% per annum)
All fats and oils	48876	60001	66178	77661	3.7	2.6
Vegetable origin	29673	42476	48003	58090	4.6	3.2
Lauric acid oils	3218	3721	3900	5312	1.8	3.6
Coconut oil	2614	2834	2854	3542	1.0	2.3
Palm kernel oil	604	887	1046	1770	4.9	7.5
Other vegetable oils	26455	38755	44103	52778	4.9	3.1
Cottonseed oil	2925	3156	3283	4044	1.0	2.5
Groundnut oil	3299	3197	3356	4253	− 0.4	2.9
Linseed oil	970	789	906	598	− 2.5	− 2.7
Olive oil	1618	1884	2002	2085	1.9	1.0
Palm oil	2420	5028	5421	10060	9.6	7.2
Rapeseed oil	2416	3553	4983	5283	4.9	4.0
Soyabean oil	6745	12990	14391	15986	8.5	2.1
Sunflower oil	3482	4913	5851	6285	4.4	2.5

Sources: FAO Production Yearbooks; IRHO (1986)

Table 3.4. Oilseeds (oil equiv.) production by region, 1985–95.

	Actual	Projected		Growth
	1985	1990	1995	1985–95
	('000 tonnes)			(% p.a.)
Industrial	17040	19686	23313	3.2
N. America	13093	16093	19147	3.2
EC-9	1900	2192	2599	3.2
Centrally Planned	5182	5806	6437	2.2
CIS	3680	4180	4693	2.2
Developing	33559	39103	46307	3.3
Asia	21640	24988	29472	3.1
Africa	3803	3670	3505	− 0.8
America	6616	8725	11380	5.6
S. Europe	1500	1720	1950	2.7
World	55781	64595	76093	3.2

Sources: FAO Production Yearbooks; IRHO (1986)

is likely to double its output, compared to 1970, and become the world's largest producer of fats and oils, after the USA.

World Trade

The fats and oils market is a highly interrelated market, characterized by substantial substitution possibilities in demand, and diverse supply sources. Vegetable oil prices tend to be highly correlated. Supplies come from both annual crops and perennial crops, from diverse geographic areas and from developing and industrial countries. Additionally, most of the oilseeds are used in the joint production of vegetable oils for human consumption and meal used for livestock. Some oils, particularly the lauric acid-containing oils (palm kernel oil and coconut oil), have significant industrial uses (World Bank, 1988).

Over the past 8 years, substantial structural changes have occurred in the market and these are expected to persist. The emergence of the European Community (EC) as a significant producer of oilseed products and protein meals, the rapid increase of productive capacity for palm oil in Malaysia and Indonesia, and expanded soyabean plantings in Latin America have brought an increased competitiveness to world markets which will be reflected in a downtrend in real prices over the long term.

In terms of market consumption, soyabean oil consumption is projected to grow in pace with total fats and oils demand at about 3% annually. Palm oil is expected to increase its share of total demand, growing at more than 6% annually. Palm oil exports to India are expected to grow. Palm kernel oil is likely to gain a larger share of the lauric oil market, increasing from 25% of the market in 1986 to 33% by 2000. Animal-source fats, cottonseed oil, and sesame seed oil are expected to lose market shares (World Bank, 1988, 1991).

Exports

In 1980, developing countries provided 39% of total world exports of vegetable oils and fats. This is predicted to rise to 51% by the 1990s. Most of this increase will come from Asia, which provided 52% of total world exports in 1980, and is expected to provide 64% by the 1990s. The major exporters of fats and oils are Argentina, Brazil, Canada, Indonesia, Malaysia, the Philippines and the USA. They are expected to account for 80% of world exports by the 1990s, compared to 75% in 1980. The major changes in exports of fats and oils expected in the 1990s are given below.

1. The main increase in exports is expected to come from Asia, particularly from palm oil, palm kernel oil and coconut oil.
2. Malaysia and Indonesia are expected to increase substantially their export

of fats and oils.

3. Other countries which are expected to increase their exports are Argentina (soyabean oil and sunflower oil), Brazil (soyabean oil) and Paraguay (soyabean oil). The Philippines may increase its output of coconut oil, if action is taken to improve coconut production.

Imports

The major changes in imports of fats and oils expected in the 1990s are given below.

1. Imports into the developing world are likely to increase from 41% of total trade in 1980 to 49% by the early 1990s.
2. The increase in imports is expected to be mainly in Asia (particularly by India, Pakistan, Korea, and China).
3. Imports are also expected to increase in Africa (particularly Algeria and Nigeria), Latin America (particularly Mexico) and the Middle East (particularly Egypt and Iran).
4. Imports by the former USSR are expected to increase substantially.
5. Imports by industrial countries are not expected to change substantially, increasing only by 0.3% per annum in the 1990s, compared with 4.9% in the 1970s.

Price Fluctuations

Vegetable oil prices are among the most volatile of all primary commodities. Copra and coconut oil prices prove to be the least stable of all the commodities covered by the World Bank (World Bank, 1988, 1991). The fluctuations in the prices of copra and coconut oil between 1950 and 2005 (predicted) are shown in Figs 3.6 and 3.7. There are considerable price risks for both producers and consumers.

The four major coconut products traded internationally are copra, coconut oil, copra meal and desiccated coconut. The price of coconut products, particularly copra, coconut oil and desiccated coconut has fluctuated over the past few decades, generally in line with other vegetable oils. The fluctuations in the monthly average price of the major coconut export products between 1985 and 1989 are shown in Fig. 3.8. This is comparable to the price fluctuations of other vegetable oils, as illustrated in Fig. 3.9. The increase in soyabean and palm oil production since the 1960s has had an overall depressing effect on international vegetable oil prices.

The relatively high level of price instability has a number of ramifications. For project evaluations, analysts must be aware of potentially large revenue swings throughout the life of the project. In development projects

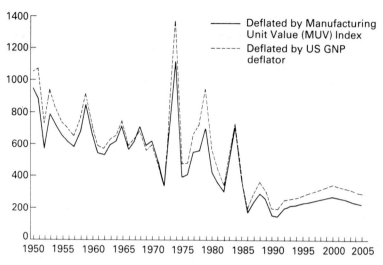

Fig. 3.6. Copra prices 1950–2005 (in US$ using 1985 as a constant). Source: World Bank, International Economics Department, 1991.

involving monocropping, smallholders remain especially vulnerable, as farmers near the subsistence level are least likely to possess the resources needed to cope with years of low price. Similarly, countries which depend heavily on vegetable oils to generate export revenue face great uncertainty (World Bank, 1988, 1991).

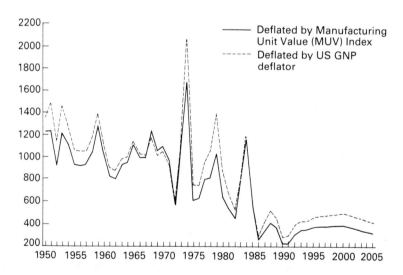

Fig. 3.7. Coconut oil prices 1950–2005 (In US$ using 1985 as a constant). Source: World Bank, International Economics Department, 1991.

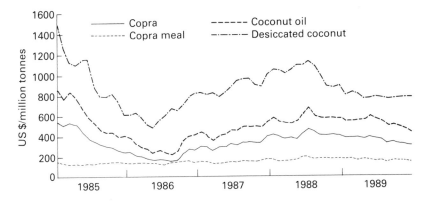

Fig. 3.8. Monthly average prices of coconut products, 1985–89. Source: Coconut Statistical Yearbook, 1989, APCC Jakarta, 1990.

Future Prospects for Coconut and Palm Kernel Oil

Although demand for industrial uses exists for most of the oilseeds, it is particularly important in the case of the lauric oils, (coconut and palm kernel oil). Historically non-edible uses of coconut oil and palm oil have accounted for more then 50% of consumption in the United States and close to 50% in Europe (World Bank, 1988).

The majority of the industrial-use lauric oils find their way into some type of soap or detergent, either directly or indirectly after the lauric oils have been converted into fatty acids and fatty alcohols. Fatty alcohols, which are used primarily in detergents, face competition from petroleum-based synthetics. In the early 1960s a large portion of the US lauric oil market was lost to

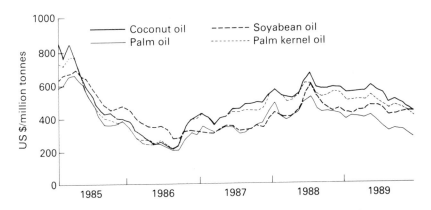

Fig. 3.9. Comparative prices of coconut oil and other vegetable oils. Source: Coconut Statistical Yearbook, 1989, APCC Jarkarta, 1990.

synthetics. Market shares now seem to have stabilized. Over the next decade, a price increase in the lauric oil market relative to petroleum is not expected to trigger major disruptions in market share. However, with limited production gains expected for coconut oil versus a large supply increase for palm kernel oil, some substitutions between the lauric oils will occur.

Currently, processing costs are higher for palm kernel oil than coconut oil for most industrial uses. In addition, coconut oil is greatly preferred as a cooking oil in most countries. Coconut oil therefore is likely to retain a preference over palm kernel oil. Producers of industrial chemicals have been showing an increasing preference for raw materials from renewable natural resources rather than petroleum-based synthetic products. There are presently few economically attractive alternatives to the lauric oils for most industrial purposes. With regard to edible uses, the general demand for vegetable oils is expected to grow by about 3% p.a. over the medium-term, and coconut oil, as a preferred cooking oil, will certainly share in that growth.

Coconut oil is expected to retain the major share of both the European and North American markets, but growth in the markets is projected to go to palm kernel oil. Coconut oil consumption is projected to range from 420,000 to 470,000 tonnes annually in Europe over the next decade, and a slightly higher demand level is expected in the United States. The consumption of palm kernel oil is expected to more than double over the next decade, reaching over 200,000 tonnes in North America and 600,000 tonnes in Europe.

Demand for coconut oil in China, India and Indonesia is expected to grow as fast as domestic production allows. However, only in the case of Indonesia is production expected to grow significantly. Demand in China is expected to remain at or below 50,000 tonnes per annum. India is expected to consume from 20,000 to 30,000 tonnes annually. Indonesia's productive capacity should allow domestic consumption of coconut oil of around 600,000 tonnes in 1988 and over 800,000 tonnes by the mid-1990s (World Bank, 1988, 1991).

Supply forecasts: lauric oils

The share of the lauric oil market held by palm kernel oil and coconut oil has changed steadily over the past years. Supplies of copra have remained stable in the Philippines, which accounted for 40–50% of world copra production in the 1960s and 1970s. Supplies have been growing in Indonesia, the second largest producer, yet domestic consumption in Indonesia has grown at an equal rate, leaving Indonesia as usually only a marginal exporter of copra and coconut oil.

At the same time, palm oil plantings have been expanded rapidly in both Malaysia and Indonesia. As a result, coconut oil's share of lauric oil production has dropped from about 85% in the early 1970s to 75% in the mid-1980s. Over the next decade its share is likely to drop below 70%.

Palm kernel oil, which can often be used as an industrial substitute for coconut oil, is extracted by crushing the kernel. With the shift into tenera oilpalms, world production of palm kernel oil declined in the late 1960s and grew slowly through the 1970s. As the number of trees increased rapidly in the late 1970s to early 1980s, production of palm kernel oil grew as well, doubling from 1976 to 1986. Most of the new plantings are now of the tenera variety of oilpalm.

Copra production remains primarily a smallholder operation in the Philippines and elsewhere, and reliable information on planted area is difficult to obtain. There has been little new planting in the Philippines over the past few decades. Many of the producing trees date back to the 1940s, and there is likely to be a decline in the productive capacity of the trees over the next 10 to 15 years. Increasing prices and greater political stability should encourage an expansion of Philippine coconut oil and copra production during the 1990s.

The World Bank approved a loan of US$121 million to the Philippines in May 1990 to enable the country to undertake a major coconut replanting programme. The purpose of this loan is to enable the Philippines to rehabilitate its coconut industry and maintain the continuity of supply of coconut oil. The loan is the largest entered into by the World Bank and the Philippines Government for many years, indicating the priority being given to coconut production by the Government of the Philippines.

Market studies by the World Bank and the Philippines Government have concluded that future demand for coconut products is good, both for industrial uses and as an edible oil, provided the Philippines can sustain end-user confidence in the continued stability of its supplies. Although the Philippines is the dominant exporter of coconut-based products, its output could fall over the coming 10–15 years because of declining yields from its ageing tree stock, unless extensive replanting is undertaken. Coconut production is a smallholder operation in the Philippines and there has been little private investment in replanting during the last decade. The Philippines provides the core of the export market, and other smaller exporters depend on the Philippines to keep the market open for coconut oil.

The share for coconut oil of both the non-edible and edible vegetable oil markets is easily sustainable, with some room for modest expansion, unless end-users become convinced that they cannot rely on a stable supply from the Philippines and other smaller exporters. In that event, the buyers will make long-term investment decisions in favour of substitute products, especially for industrial purposes. That would be a major setback for the Philippines itself and for other coconut producers. Deterioration of this industry would have important macroeconomic consequences for the Philippines (particular-

ly as regards loss of foreign exchange earnings) and would seriously under-
mine the Government's poverty alleviation efforts).

Future prospects for coconut oil

The declining availability of exportable surpluses of coconut oil in producing
countries presently prevents coconut from taking advantage of the expanding
market for vegetable oils and makes end-users apprehensive over the long
term future supply in international trade. Coconut's share of the total export
market for vegetable oils has been declining steadily. It provided 6% of the
total world market in 1986, and 5% in 1990. Approximately 70% of all
coconut production is consumed domestically and about 30% traded internationally. The current World Bank forecasts for the future of coconut oil
exports indicate the continued availability of markets for coconut oil,
provided that continuity of supply is maintained. This is an added incentive
for research efforts to address production problems in the producing
countries.

The future prospects for coconut are favourable; they depend on:

1. the growing demand for vegetable oils and fats, especially in developing
countries;
2. the increasing importance of domestic consumption of coconut in producing countries;
3. the continuing preference for the lauric acid oils (coconut and palm kernel
oil), primarily for their industrial uses in soaps and detergents.

Buyers still favour the lauric acid oils, and there is a continuing demand for
regular supplies. Coconut oil is also preferred over palm kernel oil, both for
domestic cooking purposes and in export markets for industrial uses. This
provides an opportunity for expanded coconut oil exports, if the productivity
of the crop could be improved.

Chapter four:

Current Research

'The present research efforts are not addressing adequately
the major problems of the crop . . .'

Overview

Recorded research on coconut dates back some 250 years, to the publication
Herbarium Amboinense, which gave a full account of the coconut palm in
Indonesia, including varietal descriptions (Harries, 1991). In more recent
times, coconut-producing countries have recognized the importance of
research and technical services for farmers and there are several national
research programmes in each of the coconut-growing regions of the world.
The major current programmes are listed in Table 4.1. Davis *et al.* (1987)
identified 193 scientists working on coconut, of whom 103 were in Asia, 48
in Africa, 32 in Latin America and 10 in Oceania at that time (Table 4.2).

In Asia, there are active national programmes in China, India,
Indonesia, Malaysia, the Philippines, Sri Lanka, Thailand and Vietnam. In
the South Pacific, there are several small but active programmes, notably in
Fiji, Papua New Guinea, French Polynesia, Solomon Islands, Western
Samoa and Vanuatu.

In the Americas and the Caribbean, the main research programmes are
in Brazil, Mexico, Jamaica and Trinidad. Currently, there is no breeding
programme in Mexico, despite the occurrence there of some lethal diseases
which are limiting production. There is a germplasm collection in Jamaica,
established when there was a major (and successful) international effort on
breeding for resistance to lethal yellowing disease. The breeding programme
is presently in abeyance.

In Africa, the major programmes are in Côte d'Ivoire and Tanzania. The
Marc Delorme Coconut Research Centre was established in Côte d'Ivoire in
1951. The active programme in Tanzania is receiving substantial support

Table 4.1. Illustrative national coconut research programmes.

Country	Research institutes	Programme	External collaboration past/present*
Asia			
China	Chinese Academy of Agricultural Sciences Hainan Island	Hybrid production (e.g. cold tolerance)	FAO/UNDP
India	All-India Coordinated Coconut and Arecanut Improvement Project (AICCAIP)	Hybrid breeding Germplasm collection, conservation and evaluation Pest and disease control (e.g. Kerala wilt) Tissue culture Agronomy	
	Central Plantation Crops Research Institute, Kasaragod (plus 11 other research centres)		
Indonesia	Agency for Agricultural Research and Development (AARD)		IBPGR* FAO/UNDP World Bank*
	Central Research Institute for Industrial Crops, Research Institute for Coconuts. Main stations at Menado, Pakuwon, Bone-Bone	Germplasm collection Hybrid breeding	
	Industrial Plantations (PTP/PNP) Coconut Research Program, Medan, North Sumatra	Hybrid breeding	France (IRHO)*
Malaysia	Malaysian Agricultural Research and Development Institute (MARDI)	Smallholder planting and rehabilitation	

The Philippines		
Private sector research programmes	Hybrid seed production Intercropping	World Bank* France (IRHO)*
Universities	Processing Food technology	
Philippines Coconut Authority (PCA)	Replanting/rehabilitation	FAO/UNDP Australia (ACIAR)*
Albay Research Centre	Cadang-cadang disease	Germany (GTZ)*
Davao Research Centre	Tissue culture Entomology (disease transmission)	
	Agronomy, nutrition entomology	FAO/UNDP Potash Institute France (IRHO)
Zamboanga Research Centre	Postharvest	UK (ODA)* Germany (GTZ)*
	Germplasm conservation Hybrid breeding Coconut wood utilization	FAO/UNDP New Zealand
Philippine Universities		
University of the Philippines, Los Banos (UPLB)	Embryo culture Tissue culture	
Visayas State College of Agriculture (VISCA)	Hybrid breeding Intercropping Village level processing Product development	

Table 4.1. Continued.

Country	Research institutes	Programme	External collaboration past/present*
Sri Lanka	Coconut Research Institute	Breeding and selection Soils and plant nutrition Agronomy Pest and disease control Coconut information service	World Bank* ADB IDRC* UK (ODA/NRI)*
	Coconut Board	Postharvest processing	
Thailand	Department of Agriculture, Horticultural Research Institute, Sawi Research Center	Hybrid breeding	UK (ODA) FAO/UNDP
Vietnam	Oil Crops Research Institute	Hybrid production	FAO France (IRHO)*
South Pacific			
Fiji	Ministry of Primary Industries	Hybrid breeding Pest control (biological control of rhinoceros beetle)	France (IRHO)* FAO/UNDP
French Polynesia	Coconut Research Institute	Hybrid breeding	France (IRHO)*
Papua New Guinea	Cocoa and Coconut Research Institute	Germplasm collection and conservation Hybrid breeding Pest control (*Scapanes*)	Australia (ACIAR)*

Solomon Islands	Levers Research, Yandina	Hybrid breeding Intercropping (cocoa/coconut) Tissue culture	Unilever, UK* Unifield, UK*
	Department of Agriculture	Disease surveys (viroid diseases)	Australia (ACIAR)*
Tonga	Ministry of Agriculture, Forestry and Fisheries	Rehabilitation and replanting strategies	FAO/UNDP New Zealand
Vanuatu	Oil Crops Research Station, Saraoutou	Hybrid breeding Disease control (foliar decay)	France (IRHO)* Australia (ACIAR)*
Western Samoa	Department of Agriculture	Hybrid breeding Pest control (biological control of rhinoceros beetle)	FAO/UNDP* World Bank ADB* GTZ
South Pacific Commission	Plant Protection Service	Pest and disease control	FAO/UNDP*
The Americas Brazil	Coconut Research Institute, Aracaju	Germplasm collection Food processing	
Jamaica	Coconut Industry Board	Germplasm collection and evaluation Lethal yellowing disease	USAID, UK (ODA) FAO International Council on Lethal Yellowing
Mexico	National Coconut Research Programme	Disease (lethal yellowing)	FAO/UNDP
Trinidad	National Coconut Research Programme	Red ring disease	

Table 4.1. Continued.

Country	Research institutes	Programme	External collaboration past/present*
Africa			
Côte d'Ivoire	Marc Delorme Coconut Development Centre	Germplasm collection Hybrid breeding	France (IRHO)*
Mozambique	National Coconut Development Programme	Germplasm evaluation	Portugal
Tanzania	National Coconut Development Programme	Hybrid breeding Agronomy Disease control (mycoplasmas) Biological control of insect pests Socioeconomics	World Bank* Germany (GTZ)* France (IRHO)* UK (Imperial College)*

*Existing research project supported by technical assistance, or collaborative research grants.

Table 4.2. Estimated numbers of coconut researchers by region.

Geoclimatic regions	Coconut researchers
Asia	
Sub-tropics/tropics, south Asia	15
Sub-tropics, east Asia	0
Humid tropics, south/S.E. Asia	88
Total (Asia)	**103**
Oceania	
Humid tropics **(Total Oceania)**	**10**
The Americas	
Sub-tropical South America	4
Tropics, modified by altitude or latitude	21
Humid tropics, Central America	0
Humid tropics, Caribbean	7
Total (The Americas)	**32**
Africa Sub-Saharan	
Semi-arid tropics	10
Sub-humid tropics	0
Equatorial wet tropics	4
Humid coastal tropics	10
Tropics modified by altiude or latitude	24
Total (Africa)	**48**
West Asia/North Africa	0
GRAND TOTAL	**193**

Source: modified from Davis *et al.*, 1987

from the World Bank and Germany. There are several other smaller programmes elsewhere in Africa.

Several of the major producing countries have autonomous or semi-autonomous organizations for coconut research and development. These usually have responsibility for all aspects of both pre- and postharvest research and development (Table 4.1). In most other producing countries which have a research programme, research and development activities on coconut are mainly the responsibility of government departments of agriculture. Within these, the organizational structures range from sections specializing in coconut, to situations where coconut is dealt with as part of the duties of the general agricultural research and extension workers. Research related to processing and marketing is sometimes handled by government, industrial, commerce or trade departments. In some countries, such as the Philippines, Indonesia and India, universities are also heavily involved in coconut research (FAO, 1991).

The national programmes are financed primarily from local sources, either by government grants or by a tax on producers. However, while many countries have their own research programmes, in none have they been comprehensive enough to find satisfactory solutions for the problems facing the crop. The reasons for this vary, with inadequate and/or erratic financing featuring prominently in most countries. Smallholder producers are unable to support expensive research programmes sufficiently, so reliance is usually placed on uncertain and short-term government contributions, sometimes supplemented by grants or loans from external agencies (FAO, 1991). There is substantial involvement of bilateral and multilateral development agencies in coconut research, usually related to specific projects to strengthen the work of one of the national programmes, by provision of technical expertise, finance, or both (Table 4.1).

Most coconut research programmes, particularly those in which breeding is involved, are of a long-term nature and must be continued over at least 10–20 years. This is due to the long interval between generations, the protracted period over which maturity occurs and hence experimental data need to be collected, the absence of a method of rapid vegetative multiplication, and the current need for field trials to cover large acreages which must be adequately maintained. Coconut research, therefore, requires continuity in financial support and available human resources. All too often one or both of these factors has not been available (FAO, 1991).

Many small and large-scale planting programmes have been initiated (often with external financial assistance by grants from bilateral development agencies or development bank loans). These replanting programmes require an adequate research base, so that appropriate technical decisions are made as to what is the best available planting material, and what are the accompanying management practices to allow the material to achieve its yield potential. Pest and disease control and nutritional requirements are important components of this package. An understanding of the socio-economic factors which influence farmers on the need to replant is also an important and often neglected aspect in many replanting schemes.

Scientifically, coconut is an 'orphan commodity', relative to other crops of similar importance. Most of the national coconut research programmes are seriously understaffed and underfunded. Even in the major producing countries in Asia, the national programmes are not supported in a manner commensurate with the economic and social importance of the crop to the country. The dearth of strong national coconut research programmes is a serious situation for the crop, given that it is the stated policy of many governments to increase coconut production by rehabilitating and/or replanting existing areas and planting new land, where available. The implementation of these policies will require substantial, long-term, financial investments, underpinned by a sound research base.

The key problems with the current research efforts are that most of the national programmes are not well supported financially, either by the government or by the industry; especially, they lack continuity of funding from national and external sources; they lack sufficient appropriately trained staff and suitable facilities; they are not addressing adequately the major problems facing the crop; they are not producing sufficient substantive results directly relevant to smallholders; nowhere are the needs of the crop world-wide being addressed; and there are presently no means by which small producing countries, which are unable to mount their own research efforts, are able to gain access to new technologies, including higher-yielding varieties.

There are several national programmes which could contribute substantially to any international initiative on coconut research, if suitably supported. A brief description of the main national coconut research programmes in different geographic regions is given below.

Asia

China

There is a small national coconut research programme in China on Hainan Island at latitude 20°N. A seed garden is producing hybrids (Malayan Dwarf × West African Tall, and Malayan Dwarf × Hainan Tall). The major problems are cold and drought. Hainan Island is close to the northern marginal limit for coconut palms, which only produce nuts there for 4 months of the year. It is thus a site of particular interest to coconut breeders for screening cold-tolerant material.

India

Coconut research in India began in 1916 with the establishment of the Central Plantation Crops Research Institute at Kasaragod. The silver jubilee of the institute was celebrated with a major seminar in 1991. The institute is now part of a national programme, the All-India Coordinated Coconut and Arecanut Improvement Project (AICCAIP), which is responsible for coordinating the activities of 12 research centres located throughout the major coconut-growing regions.

The objectives of the national programme are:

1. collecting, conservation, documentation, and evaluation of germplasm;
2. evaluation of hybrids and high-yielding cultivars;
3. standardization of research techniques for various agroclimatic regions;

4. development of suitable multiple cropping and intercropping systems;

5. development of efficient pest and disease management systems.

There is a germplasm collection at the Central Plantation Crops Research Institute, containing some 60 exotic cultivars and 30 local varieties. A quarantine station in the Andaman Islands maintains some exotic coconut collections.

Although extensive nutritional trials have been conducted with mono-cropped local varieties, little is known of the nutritional needs of locally produced or introduced hybrids in a monocrop or intercrop situation. Drought is a major problem in Indian coconut production.

The most damaging coconut disease in India is Kerala wilt disease. Losses due to the disease have been estimated at approximately 340 million nuts per year (equivalent to 40,000 tonnes of coconut oil). A mycoplasma has been shown to be the casual agent. A possible vector for the disease has been identified.

Indonesia

Coconut research in Indonesia commenced in the early 1900s. The first research station was established in 1930 at Menado. Most research now comes under the direction of the Agency for Agricultural Research and Development (AARD). The responsible institute within AARD is the Central Research Institute for Industrial Crops. Its research institute for coconuts is located near Menado in northeast Sulawesi, with a sub-station at Pakuwon (West Java), and 11 other stations. The Bone-Bone station has a substantial germplasm collection. AARD has collaborated with FAO/UNDP, the World Bank, IBPGR, and France (IRHO) in its coconut research and development programmes.

The state-owned industrial plantation group which supports nucleus estates and associated coconut smallholdings also funds a coconut research programme through the Pusat Penelitian Kelapa (PPK). This programme was established in 1982 at Bandar Kuala near Medan in North Sumatra. The PPK has received technical assistance from France, which has provided two IRHO scientists (an agronomist and an entomologist) to assist a team of ten Indonesian scientists.

Malaysia

Coconut research is undertaken by the Malaysian Agricultural Research and Development Institute (MARDI). MARDI's coconut research programme is concerned with the research needs of the smallholder planting and rehabilitation progamme. Private sector research interests have contributed to research on hybrid seed production, and cocoa and coconut intercropping.

Malaysian universities have done some interesting work on coconut processing and food technology, notably the development of a process to produce 'instant' coconut cream powder, which can be reconstituted to coconut cream by the addition of water. This product is now marketed internationally.

The Philippines

Research on coconuts has been undertaken by various organizations in the Philippines since 1908. The research was largely fragmentary and discontinuous, until the Philippines Coconut Authority (PCA) was established in 1973. Research is now undertaken by the PCA (a statutory authority under the Minister of Agriculture), and several universities.

Much of the coconut area in the Philippines is occupied by senile palms. In May 1990, the World Bank approved a US$120 million loan to the Philippines to enable it to embark on an ambitious coconut rehabilitation and replanting scheme. This programme will require a sound research and extension base to ensure its success.

Philippines Coconut Authority

The PCA research programme was established with support from the Government of the Philippines, and FAO/UNDP through a Coconut Research and Development Project, from 1971 to 1982.

There are three principal PCA research stations, which employ a total of approximately 40 scientists, plus support staff. The research activities of each station are described below.

Albay Research Centre, Southern Luzon   Albay Research Centre has concentrated on cadang-cadang disease, which kills millions of coconut trees each year in the Philippines. It has been estimated to cost the Philippines USS$16 million per year in lost production. The causal agent of the disease has been shown to be a viroid. Collaborative research has been undertaken by scientists from PCA and the University of Adelaide, Australia over the past decade, with financial support from FAO/UNDP, and the Australian Centre for International Agricultural Research (ACIAR), to identify the causal agent and establish effective control programmes (Hanold and Randles, 1991a). A resistance screening method has been devised, and germplasm is being screened for tolerance to cadang-cadang. A field assay technique has also been developed and an extensive field survey is under way to determine the geographic spread of the disease. Cadang-cadang appears to be more widely distributed in the Albay area and the neighbouring regions than has been recognized previously. The Albay area was severely damaged by a typhoon in early 1988 and much coconut rehabilitation is required. Tissue culture research has been initiated at Albay Research Centre, with support from Germany via GTZ.

Davao Research Centre, Southern Mindanao   Research here is concerned primarily with agronomy, particularly the study of nutritional deficiencies such as nitrogen and chlorine; and entomology, particularly biological control of rhinoceros beetle and other coconut pests.

Zamboanga Research Centre, South-West Mindanao   Zamboanga is the principal PCA breeding and selection station. A germplasm collection with 74 accessions has been assembled. The initial collections were sponsored by PCA and the International Board for Plant Genetic Resources (IBPGR). The maintenance of the collection after the completion of the IBPGR support is a continuing problem. Much of the collected material has not yet been comprehensively described and evaluated.

In regard to the breeding and evaluation, both local and exotic hybrids are being assessed in a national hybrid evaluation trial in a programme supported by PCA and the Philippines Council for Agricultural Research and Rural Development (PCARRD). Emphasis is on the identification of suitable locally-adapted hybrids for the different regions. Some promising local hybrids have been identified.

Zamboanga Research Centre is also the site of research on coconut wood utilization. Experiments over several years, conducted by PCA scientists, in cooperation with FAO and New Zealand forestry specialists, have shown that coconut wood can be used in the construction of houses, as well as furniture and specialty items such as parquet flooring. This research was designed to find uses for old palms, so that palms felled in replanting programmes would yield an income to the farmer and their removal would reduce the risk of rhinoceros beetle infection in new plantings. A cash return on the old trees alters the economics of replanting schemes, and makes them more attractive to smallholders. This work has important implications for all coconut-growing countries.

Philippines universities   Research on coconut is also conducted at several universities, particularly the University of the Philippines at Los Banos (UPLB) and the Visayas State College of Agriculture (VISCA). UPLB has a long history of tissue culture research. Dr de Guzman first successfully cultured coconut embryos there, to allow the propagation of Makapuno nuts, a coconut variety with jelly-like endosperm which is a local delicacy.

Research at VISCA is concerned with the production and evaluation of hybrids suitable for local conditions. Research is also conducted on inter-cropping, in order to recommend the best replanting options to farmers. Village-level processing methods are also being developed at VISCA and elsewhere in the Philippines, in order to make the best use of the total coconut resource. Considerable work has been done on coconut product development, including various food products.

Sri Lanka

The Coconut Research Institute of Lunuwila was established in 1929. Results in its early years contributed to coconut breeding and selection, nutritional requirements, and pest and disease control. A seed garden was established in 1955 for the production of improved tall material and for dwarf × tall hybrid seed. The infrastructure has been upgraded regularly with outside financing. The Asian Development Bank provided funds for improving laboratory facilities and for establishing a seed garden, and the European Community provided assistance and for the establishment of the third seed garden. The Sri Lankan government is now seeking assistance from the Asian Development Bank for further improvement of the coconut industry. Currently, the World Bank is providing assistance for infrastructure development and for manpower training. There are about 34 scientists on the staff, 15 of whom have received postgraduate training with a further seven receiving postgraduate training overseas at present.

The six research divisions of the Institute are concerned with: (i) soils and plant nutrition; (ii) genetics and plant breeding; (iii) agronomy and agricultural economics; (iv) crop protection; (v) plant physiology; and (vi) tissue culture. There are supporting divisions for biodiversity, estates and publications.

The Institute maintains a coconut information service. This service has received external support from the International Development Research Centre (IDRC). It is now linked more closely with the APCC Secretariat in Jakarta in the provision of coconut information services in a new IDRC-supported project.

The Coconut Board in Sri Lanka has a particular interest in improving postharvest processing. The UK, through the Natural Resources Institute (NRI), has supported postharvest technology projects in Sri Lanka for several years. These are concerned primarily with the development of improved methods of processing copra.

Thailand

Coconut research is conducted at the Sawi Research Centre in southern Thailand, part of the Horticultural Research Institute of the Department of Agriculture. The United Kingdom, through ODA, provided technical and financial support for the programme for several years during the 1980s. A seed garden at Kanthuli has produced hybrid seed nuts for distribution to smallholders. The preferred hybrid is the Thai Tall × West African Tall, which produces a large nut. The main market in Thailand is for whole nuts, and nut size is a critical factor.

Food technology research at Kasetsaart University has been successful in developing processes to allow the inclusion of dried coconut milk in 'instant' curries, which can be reconstituted by the addition of water. These products are now exported to the USA, and elsewhere.

Vietnam

There is a research institute for oil crops in Vietnam, at which some coconut hybrid evaluation has been conducted. A hybrid seed garden has been established. FAO has provided technical assistance, by the provision of a coconut breeder. France also provides technical assistance through IRHO.

South Pacific

There are several small research programmes in the South Pacific, notably in Fiji, French Polynesia, Papua New Guinea, Solomon Islands, Tonga, Vanuatu and Western Samoa. These are described briefly by Foale (1987). Technical assistance is provided by France through IRHO to the programmes in Fiji, French Polynesia and Vanuatu. FAO has also provided technical assistance to the programmes in Western Samoa and Tonga.

Collaborative research has been established with the sponsorship of ACIAR amongst the programmes in Papua New Guinea, the Solomon Islands, Vanuatu and several Australian institutions (CSIRO; Victorian Department of Agriculture and Rural Affairs; University of Adelaide; University of Melbourne). A South Pacific regional coconut network was established by the Institute for Research, Extension and Training in Agriculture (IRETA) of the University of the South Pacific, in cooperation with the South Pacific Commission in the 1980s, but has since lapsed.

Fiji

The first hybrid coconut palms ever produced by artificial pollination were grown in Fiji by Marechal in 1928. He crossed an introduced Malayan Red Dwarf with Fijian Green Dwarf (Niu Leka). It was the variability shown by the F_2 progenies of the Fijian hybrid material when growing at Yandina in the Solomon Islands, that demonstrated the potential benefits from coconut breeding.

After the initial crossing programme in Fiji, there have been several attempts to establish a hybrid breeding programme there, but none of these has resulted in any substantial production of hybrid material for use in replanting schemes. A new programme commenced in 1985, with technical assistance being provided by IRHO, with the aim of producing hybrids to replace the large proportion of senile palms in Fiji. This programme, based on Taveuni, has made significant progress towards producing hybrid seed on a large scale.

A successful biological control programme for rhinoceros beetle, using baculovirus, has been conducted in Fiji and several other Pacific islands since the 1960s, with support from UNDP and FAO. The beetle is controlled by an introduced baculovirus. Regular releases of the virus are still made, to maintain the effectiveness of the biological control (Waterhouse and Norris, 1987).

French Polynesia

A small hybrid breeding programme is conducted under the auspices of IRHO. The pioneering work on the nutrition of coconut under atoll conditions, which was conducted in French Polynesia during the 1960s was a significant contribution to knowledge in this area.

Papua New Guinea

Coconut research has been conducted in Papua New Guinea for many years, with support from both the public and private sectors. In 1986 the Papua New Guinea government established the Cocoa and Coconut Research Institute (CCRI) in Rabaul. The institute is largely funded by a levy on growers. It is managed by a Board of Directors representing producers, exporters, and the government. The institute employs several graduate staff. These include a full-time coconut breeder and agronomist, and a pathologist and entomologist who share their time between coconut and cocoa.

The major initial work at the institute has been to establish a programme for germplasm collection within Papua New Guinea, arrange for the introduction of potentially useful material, and establish a hybrid breeding programme. Pest control is also important, since pests, particularly *Scapanes* beetle, are a major threat to production. Diseases are less serious than pests at present, but these need to be monitored carefully. Embryo culture techniques are being established in PNG in collaboration with Australian scientists. These techniques, when combined with disease indexing, should allow the safe introduction of germplasm (Hanold and Randles, 1991b).

Solomon Islands

There is a long tradition of coconut research in the Solomon Islands, dating back to 1952, when a research programme was established at Yandina by

Levers (a subsidiary of Unilever). In 1960, a joint coconut research scheme was established between the Solomon Islands Government and Levers. This scheme operated until 1975, when the research programme became the sole responsibility of Levers. A brief description of the past research at Yandina and the current research programme is given by Foale (1987).

There is presently one coconut breeder stationed at Yandina. Current research includes the continued production and evaluation of hybrids; intercropping trials to determine optimum palm spacing to suit intercropping with cocoa, and the youngest age at which coconut may be underplanted with cocoa. Yandina also provides coconut tissues to the Unifield laboratories in Britain who are working on somatic embryogenesis in coconut. A single palm planted in 1985 at Yandina was produced from coconut callus culture in the UK, one of the few such palms growing in the field anywhere in the world.

Vanuatu

A coconut research programme was established in Vanuatu in 1962, at the Saraoutou Oil Crops Research Station on the island of Espiritu Santo. It is managed by IRHO on behalf of the Government of Vanuatu. IRHO has four scientists stationed at the Saraoutou research station. The work at the station has been described in detail by Calvez *et al.* (1985).

The station has established a hybrid breeding programme, and a hybrid seed and seedling production system. The system could be used as a model for nursery management elsewhere. The station has produced a local hybrid with higher yield potential than the local tall variety and tolerance to foliar decay disease.

Saraoutou is the best equipped and staffed coconut research facility in the South Pacific. The major constraint to the station being the regional centre for coconut research is the presence of foliar decay disease. This disease affects most introduced varieties, while the local tall and dwarf trees are tolerant. The insect vector of the disease is a leaf hopper (*Myndus taffini*). All hybrids developed for use in Vanuatu will have to have tolerance to the disease. Germplasm from Vanuatu cannot be distributed with safety elsewhere, unless it goes through embryo culture and disease indexing, because of the risk that the virus may be seed-borne.

Research by scientists in Vanuatu, IRHO, France, and the University of Adelaide, Australia (with support from ACIAR), has shown that foliar decay disease is caused by a small, circular, single-stranded DNA virus (Randles *et al.*, 1987; Rohde *et al.*, 1990). A diagnostic probe has been developed to allow field surveys to be made of the distribution of the disease in Vanuatu and elsewhere, and for use in other epidemiological studies and for disease indexing. A diagnostic test based on the use of monoclonal antibodies is also being developed (Randles, personal communication).

The Americas

Brazil

Brazil aims to become self-sufficient in lauric oils, and has decided to expand its coconut production. To this end, Brazil has a national coconut research centre at Aracaju, with a germplasm collection. It has embarked on a hybrid breeding programme.

Brazil makes extensive use of coconut milk, coconut cream and coconut flakes in cooking. The production from 50,000 ha of coconuts is consumed as food within Brazil. The processing technology developed in Brazil could be of considerable interest to other countries wishing to develop these items as value-added products for local sale or export. Brazil also uses lauric acid extracted from coconut oil in its oleochemical industry, and this demand is increasing.

Jamaica

There is a long-standing research programme in Jamaica, where the Coconut Industry Board began its research department in the 1960s. Ths programme has been funded in the past by a producers' levy and from international and bilateral sources, particularly the UK. Most of the external assistance has ceased in recent years.

A germplasm collection was established as part of a programme on lethal yellowing disease, first reported from the Caribbean. The disease has been shown to be caused by a mycoplasma-like organism (MLO). Resistance was identified in some introduced germplasm, especially Malayan Red Dwarf. A hybrid with the Panama Tall, known as Maypan, has been widely planted.

USA: Florida

The University of Florida made a major commitment to research on coconut in 1971, when lethal yellowing disease was discovered on the US mainland. The Ford Lauderdale Research and Education Center undertook an intensive research program on lethal yellowing for the next 10 years. Three plant pathologists, two entomologists, a horticulturalist and an electron microscopist were assigned to the programme in Florida. Another scientist was seconded to the Coconut Industry Board in Jamaica. There was close cooperation between the research programmes in Florida and Jamaica, and several aspects of the disease (including its causal agent and vector) were elucidated (Howard, personal communication).

The Fort Lauderdale centre has maintained its interest in coconut research over the past decade, although at a less intensive level than at the

height of the lethal yellowing epidemic. Currently there are two horticultural-
ists, a plant pathologist, and an entomologist interested in palms. There are
major projects on three of the most important pests of coconut palms in the
Americas: lethal yellowing, red ring and coconut mite (*Accria guerronis*). The
research is conducted in collaboration with scientists in Jamaica, Trinidad,
Puerto Rico and Costa Rica. The centre also provides training for scientists
from other coconut-growing countries in the biology, taxonomy and horticulture of palms.

Centre scientists also provide advisory services to other countries especially in regard to the diagnosis of coconut pests and diseases in the Americas.
The centre also hosted the International Council on Lethal Yellowing for
several years in the 1970s (Howard, personal communication).

Other countries

There are small national research programmes in Mexico, Panama, Surinam,
Trinidad and Venezuela. Lethal yellowing disease is threatening the expansion of the coconut industry in Mexico. Little coconut research is being done
in Colombia, Ecuador, and Peru, where priority is given to oilpalm.

Africa

The current status of coconut research in Africa was reviewed at a BURO-
TROP seminar held in Tanzania in February 1991. A synopsis of some of the
programmes discussed at the seminar is given below.

Benin

Research on coconut is carried out at one of ten national research stations,
on: (i) seed production; (ii) precocity of hybrid PB 121; (iii) coconut-based
cropping systems; (iv) nutrition; (v) varietal yields (BUROTROP, 1991a).

The major current constraint is the lack of seed for replanting programmes. The major pest problem is the mite, *Eriophyes*. The lethal
mycoplasma-like disease common further along the West African coast has
not yet been sighted in Benin, but its movement is being monitored. Postharvest technology is concerned particularly with the development of new
food products.

Cape Verde

A small programme on the evaluation of coconut varieties is being conducted
in collaboration with IRHO. Mites are a problem on some of the newly
introduced varieties (BUROTROP, 1991a).

Côte d'Ivoire

Although only a relatively small coconut producer, the Côte d'Ivoire has a substantial coconut research institute, the Marc Delorme Coconut Research Center, established in 1951 by IRHO. The research priorities are: (i) lethal yellowing disease; (ii) intensification of cropping systems; (iii) selection of highly productive hybrids; and (iv) value-added products.

The major emphasis has been on germplasm collection and hybrid breeding and evaluation. The comprehensive germplasm collection contains 52 natural varieties (34 talls, 18 dwarfs) and 74 hybrids. Embryo culture techniques are being used to facilitate germplasm collecting and exchange. There has also been a significant research effort on coconut nutrition.

The hybrid PB 121 (Malaysian Yellow Dwarf × West African Tall) was bred in the Côte d'Ivoire. It is now grown commercially in approximately 40 countries. Some other promising hybrids in experimental trials are PB 132 (Malaysian Red Dwarf × Polynesian Tall) and PB 213 (Rennell Tall × West African Tall). There are about 500 ha of hybrid trials at the centre (IHRO, 1986).

Ghana

The major constraint to coconut production in Ghana is a lethal disease, Cape St Paul Wilt, which is killing many palms along the coast. A World Bank loan is being sought to establish a research programme on the disease (BUROTROP, 1991a). It is proposed that coconut research will be included within the Oil Palm Research Institute, with at least five scientists working on coconut.

Nigeria

Nigeria gives priority to research on oil palm in preference to coconut, which is consistent with the relative importance of the crops in the country. Nigeria's modest national coconut programme (which commenced in 1964) is concerned primarily with: (i) introduction and evaluation of germplasm; (ii) the lethal wilt diseases in West Africa; (iii) sustainable coconut-based intercropping systems; and (iv) expansion of seed gardens. The major constraint to the expansion of coconut planting is the lack of sufficient seed (BUROTROP, 1991a).

Tanzania

Research is conducted by the National Coconut Development Program (NCDP). The programme is closely linked to a World Bank development project for the rehabilitation of the coconut lands in Tanzania. Technical assistance has also been provided by France (IRHO), Germany (GTZ), and the UK (Imperial College, University of London). Four scientists sponsored by these institutions have been based in Tanzania over several years. The programme includes research on breeding, especially for resistance to the lethal mycoplasma-like disease; biological control of insect pests, particularly *Oryctes monoceros* and *Pseudotheraptus wayi*; agronomy; and socioeconomic constraints affecting smallholders.

Priority research areas for Tanzania are the development of: (i) diagnostic techniques for mycoplasma-like organisms; (ii) evaluation techniques for drought tolerance; (iii) biochemical techniques to distinguish different coconut types and (iv) farming systems (BUROTROP, 1991a).

International and Regional Research Activities

There are several international and regional organizations which support significant research activities on coconut. An illustrative list of projects with which various agencies have been involved with national programmes is contained in Table 4.1. The major agencies active in coconut research are:

- Asian and Pacific Coconut Community (APCC);
- Australian Centre for International Agricultural Research (ACIAR);
- Bureau for the Development of Research on Tropical Perennial Oil Crops (BUROTROP);
- Food and Agriculture Organization of the United Nations (FAO);
- Institut de Recherches pour les Huiles et Oleagineux (IRHO), France;
- International Board for Plant Genetic Resources (IBPGR);
- International Council for Research in Agroforestry (ICRAF);
- South Pacific Commission (SPC);
- Transnational companies;
- United Kingdom Overseas Development Administration (ODA);
- United States Agency for International Development (USAID);
- United States Department of Agriculture (USDA);
- World Bank.

Asian and Pacific Coconut Community

The APCC was established in 1969 under the auspices of the Economic and Social Commission for Asia and the Pacific (ESCAP), to foster cooperation amongst producing countries in all aspects of coconut production, processing and marketing. The APCC is an intergovernmental organization, currently with a membership of 14 countries. Its present members are the Federated States of Micronesia, Fiji, India, Indonesia, Malaysia, Papua New Guinea, the Philippines, Solomon Islands, Sri Lanka, Thailand, Vanuatu, Vietnam and Western Samoa. The Republic of Palau is an associate member. The APCC has a permanent secretariat based in Jakarta and a network of liaison offices in its member countries. The member countries of the APCC account for approximately 85% of world production of coconut.

The APCC meets at least once in each calendar year, and the member countries are represented by plenipotentiary delegates. The technical arm of APCC is the Permanent Panel on Coconut Technology (COCOTECH). It meets annually, and sets guidelines for APCC's work programme. Regular COCOTECH meetings have permitted member countries to exchange information and assess regional priorities for research and development in the areas of production, processing and marketing of coconut products. Each annual COCOTECH meeting takes a different research theme. APCC also provides a forum for seminars and workshops of particular interest to the industry.

In 1990 the APCC commenced a comprehensive information and documentation project, with financial support from the International Development Research Centre (IDRC) of Canada. This project enables APCC to expand its excellent publications and information service, which currently produces the annual Coconut Statistical Yearbook, a bimonthly newsletter 'Cocomunity', a biannual technical journal (CORD), the proceedings of specialized workshops and seminars, and a series of directories on exporters, importers, manufacturers of equipment and research workers in the coconut sector (Punchihewa, 1991).

Australian Centre for International Agricultural Research

ACIAR supports collaborative research among Australian scientists and their colleages at institutions overseas, especially in Asia and the Pacific. Soon after its establishment in 1982, ACIAR commenced support for coconut research with a primary focus on activities in the South Pacific (Wright and Persley, 1988). ACIAR is supporting a coconut improvement project involving the Cocoa and Coconut Research Institute (CCRI) in Papua New Guinea, the Victorian Department of Agriculture, and the CSIRO. The project includes germplasm collecting within Papua New Guinea, the development of embryo culture techniques to facilitate germplasm exchange and some tissue culture research.

ACIAR is supporting collaborative research on coconut virus and viroid diseases by scientists at the University of Adelaide, the Philippines Coconut Authority, the Saraoutou Oil Crops Research Station, Vanuatu, and IRHO. This research is concerned with cadang-cadang disease in the Philippines and the occurrence of viruses and viroids in coconut palms in the South Pacific (Rohde *et al.*, 1990). The research has shown that foliar decay disease in Vanuatu is caused by a virus. It has also shown that viroids similar to that which causes cadang-cadang are widely distributed in the Pacific Islands and southeast Asia (Hanold and Randles, 1991b).

The potential for the biological control of coconut pests in the Pacific Islands was reviewed by Waterhouse and Norris (1987). Within the socio-economic area, ACIAR-supported research in Papua New Guinea has collated data on tree crop cultivation including coconut. In addition, ACIAR has been examining market prospects for coconut products (ACIAR, 1988).

Bureau for the Development of Research on Tropical Perennial Oil Crops

BUROTROP was established in 1990 under the sponsorship of the European Community (EC). It aims to coordinate support for research on coconut and oilpalm, and to improve linkages between research and related development projects in producing countries.

BUROTROP is the result of an extensive review commissioned by the EC on the need for the creation of a research organization for tropical perennial oil crops. The study documents describe European-based coconut research activities to the mid-1980s and suggests options for future support by the EC (IRHO, 1986).

BUROTROP consists of an Executive Committee of 15 members, currently comprised of seven members from producing countries, seven from European member countries, and one member from the EC. It is supported by a small secretariat (a Director and a Secretary/Translator) based in Paris. Its structure and mandate is such that it intends to coordinate research on oilpalm and coconut, but not to conduct research itself. It has been established for an initial period of 3 years, with funding being provided primarily by the EC (BUROTROP, 1991b).

The immediate objectives of BUROTROP are to:

1. organize the exchange of information and experience amongst coconut and oilpalm research institutes;
2. create a database on tropical perennial oil crops in conjunction with specialized organizations;
3. identify production constraints and research and development needs;
4. reinforce research and development projects within national programmes and foster existing and new networks;
5. analyse research and development staff training requirements and identify organizations capable of providing training;
6. organize seminars, conferences and workshops to encourage the creation of networks;
7. seek funding for network or national activities;
8. set up liaison mechanisms and publish activity reports and bulletins.

The initial priority areas for BUROTROP are Africa, and oilpalm. In February 1990, it sponsored a seminar on coconut research needs in Africa

(BUROTROP, 1991a). A similar seminar is planned for the Americas in 1992.

Food and Agriculture Organization of the United Nations

FAO, with financial support either from its own regular budget, or from UNDP, has supported several coconut research and development projects. The major projects have been in the Philippines and Indonesia. In addition, FAO has assisted a number of small-scale national activities on specific aspects of research and development. These have included projects in Asia (China, Bangladesh, Pakistan, Thailand, Vietnam), the Pacific Islands (Marshall Islands, Tonga, Western Samoa), Africa (Guinea–Bissau), Latin America (Equador, Mexico), and the Caribbean (Dominican Republic).

A global forum for information exchange and discussion of research and development priorities was the FAO Technical Working Party on Coconut Production, Protection and Processing, which held meetings every 4 years during the 1960s and 1970s. In the mid-1960s, the Technical Working Party gave rise to a Coconut Breeders Consultative Committee which assisted FAO in producing (until 1979) an annual report for interchange of information among coconut breeders (Harries, personal communication). FAO was also a member of the International Council for Lethal Yellowing, which was active for several years in the 1970s.

An Asia/Pacific regional coconut project provided training and workshops on several aspects of coconut production in the mid-1980s (e.g. biological control of *Oryctes*, genetic improvement, integrated pest management, coconut replanting, diseases of unknown cause, nutritional deficiencies, coconut-based farming systems, processing, intercropping, and farmers' receptivity to new coconut techniques) (FAO, 1991).

Institut de Recherches pour les Huiles et Oleagineux

IRHO is one of the member institutes of the French Centre de Cooperation Internationale en Recherche Agronomique pour le Development (CIRAD). IRHO operates a multicountry programme on coconut research, which has been in operation for some 30 years. Its major breeding activities have been at the Marc Delorme research station in the Côte d'Ivoire and the Saraoutou station in Vanuatu. It also has laboratories in Montpellier, France. IRHO has staff stationed with national programmes in several countries, currently including Brazil, Fiji, Indonesia, Papua New Guinea and the Philippines. IRHO's research activities have been concerned primarily with addressing the needs of planting programmes, particularly by the production and evaluation of new hybrids, and establishing their nutritional requirements. IRHO's best available hybrid has a copra yield potential of about 6.5 tonnes/ha (compared to the current world average of 0.5–1.0 tonnes/ha).

The Marc Delorme Research Centre in Côte d'Ivoire has a substantial germplasm collection and a number of promising hybrids, the result of some 40 years of research. Any internationally supported coconut improvement programme will need to find a way to build on the substantial financial and scientific investment at this station to enable it to continue to contribute to coconut research internationally.

International Board for Plant Genetic Resources

IBPGR has been conducting and commissioning coconut research since the early 1980s. Coconut has been given a high priority by IBPGR for germplasm collecting, especially in southeast Asia and the Pacific. Between 1980 and 1985 some 320 accessions were collected and planted in different field genebanks, mainly in the Philippines, Indonesia, Malaysia and India. A small collecting mission was also sponsored in Mexico. In 1981, IBPGR published a directory of coconut germplasm collections.

The problems encountered during collecting and exchange of germplasm, mainly caused by the large size of the seed, led IBPGR to commission research on *in vitro* collecting and conservation of coconut, with the following objectives:

1. to develop *in vitro* methods for collecting and culturing mature zygotic embryos;
2. to raise embryos in culture and transfer them to independent *in vitro* growth;
3. to investigate the conditions required to culture immature embryos; and
4. to cryopreserve zygotic embryos.

The most promising prospects for the long-term storage of coconut germplasm by cryopreservation appear to lie with immature zygotic embryos and, perhaps, somatic embryos. This research was commissioned by IBPGR, and undertaken by IRHO and ORSTOM. Recent results suggest that cryopreservation of immature, zygotic embryos is a viable means for *in vitro* storage of coconut germplasm. This could be used as a complement to field-based germplasm collections.

IBPGR organized a training course in the Philippines on the collecting, conservation and characterization of coconut genetic resources where participants from several countries learned how to apply the latest techniques in *in vitro* collecting, characterization and conservation. These methods are now feasible for use in all coconut-producing countries.

In 1991, IBPGR began fostering the establishment of an International Coconut Genetic Resources Network, in response to TAC's recommendations on the need for additional research on coconut, and the priority of the genetic resources activities. The scope of the new network is discussed in Chapter five.

International Council for Research on Agroforestry

ICRAF included descriptions of several coconut-based agroforestry systems in its inventory on agroforestry systems in the tropics (Nair, 1989). These systems are described in Chapter two. The ICRAF study also identified future research needs to enable the development of more sustainable coconut-based systems which are still economically attractive to small-holders. Now that ICRAF has become a member of the CGIAR it is expanding its research activities in Asia. In the future these may include research on coconut-based agroforestry systems in Asia.

South Pacific Commission

SPC is an intergovernmental organization which assists member countries by information exchange, technical meetings and training. In the 1960s, SPC initiated a joint SPC/UNDP project for research on rhinoceros beetle control in Western Samoa. Its success provided the basis for effective biological control of rhinoceros beetle throughout the Pacific region. More recently, the SPC Plant Protection Service monitors the biological control of rhinoceros beetle, surveys the incidence of other coconut pests and diseases, and makes recommendations on their control for all 22 SPC member countries.

Transnational companies

The main transnational companies involved in coconut production and processing are Proctor and Gamble, Unilever, and Harrison, Crossfield and Flemings. The last two companies operate a joint research facility in the UK ('Unifield'). There is some applied research, especially in agronomy, nutrition and varietal testing, on large estates in producing countries. There is also some laboratory-based research by private companies (notably on coconut tissue culture by the Unifield laboratories in the UK). The primary aim is to develop a clonal propagation sysem for coconut similar to that developed for oilpalm.

Most private sector companies are involved in processing and utilizing coconut commodities in both producing and importing countries; they conduct some research aimed at improving quality, reducing cost, or widening the market for their particular products (FAO, 1991).

Overseas Development Agency, UK

During the past two decades, ODA has supported several research projects on coconut, (e.g. on coconut improvement in Thailand and lethal yellowing in Jamaica). It has also provided long-term support for research in the UK on tissue culture at Wye College, and research on coconut diseases at the International Mycological Institute (IMI) and the Natural Resources Institute (NRI).

NRI has also had a long-standing involvement in postharvest research and development activities. Current NRI projects in Sri Lanka and the Philippines are concerned with the development of improved coconut processing technology. ODA is an active member of BUROTROP, and currently provides the Chair of the Executive Committee.

United States Agency for International Development

USAID supported research on lethal yellowing in the Americas for several years. This included support for an entomologist with the Coconut Industry Board in Jamaica (1976–80), and collaborative research between the University of Florida Fort Lauderdale Research and Education Center and the Jamaica Coconut Industry Board (1986–88) (Howard, personal communication).

United States Department of Agriculture

USDSA has supported research on coconut pests and diseases in Florida, primarily on lethal yellowing disease, and coconut mites. Currently, USDA is supporting research at the University of Florida Fort Lauderdale Research and Education Center on DNA probes for the diagnosis of lethal yellowing (Howard, personal communication).

World Bank

The World Bank supports coconut research programmes primarily through the research components within national coconut and/or tree crops development projects. Current Bank projects which include a component for strengthening the national coconut research programme are located in Indonesia, the Philippines, Sri Lanka and Tanzania. A national agricultural research project supported by the World Bank in India has recently included support for research on a rapid propagation system and Kerala wilt disease on coconut. The World Bank is also one of the co-sponsors of the CGIAR, and has supported the inclusion of coconut as one of the CGIAR commodities.

Plate 1
Coconut climber in south India
(photo: G. Toomey)

Plate 2
Sri Lankan tall coconut type
(photo: M.A. Foale)

Plate 3
Dwarf coconut, Philippines
(photo: G.J. Persley)

Plate 4
Dwarf coconuts, Cook Islands
(photo: M.A. Foale)

Plate 5
Coconut fruit characteristics
(Malaysian and yellow dwarfs),
Ronatonga, Cook Islands
(photo: M.A. Foale)

Plate 6
Copra making, Tonga
(photo: G.J. Persley)

Plate 7
Coconut wood utilization,
Philippines
(photo: G.J. Persley)

Plate 8
Coconuts underplanted with
food/beverage crops, Vanuatu
(photo: G.J. Persley)

Plate 9
Coconut/cocoa intercropping, Papua
New Guinea
(photo: M.A. Foale)

Plate 10
Monocropped coconut palms in the Philippines
(photo: G.J. Persley)

Plate 11
Clonal coconut palm grown from
tissue culture in the UK, growing at
Yandina, Solomon Islands
(photo: M.A. Foale)

Plate 12
Coconut nursery, Zamboanga Research Station,
Philippines
(photo: G.J. Persley)

Plate 13
Symptoms of foliar decay disease,
Vanuatu
(photo: J. Raff)

Plate 14
Screening for tolerance to foliar decay, Vanuatu
(photo: M.A. Foale)

Plate 15
Scapanes beetle damage, Keravat,
Papua New Guinea
(photo: M.A. Foale)

Plate 16
Palm weevil (adult and pupa), Keravat, Papua New
Guinea
(photo: M.A. Foale)

Chapter five:
International Research Priorities

A number of activities can be addressed realistically only
through an international programme . . .

Beneficiaries of Coconut Research

Coconut was one of 12 commodities analysed by Davis *et al.* (1987), in a
model for the assessment of agricultural research priorities. On the grounds
of efficiency, investments in coconut research are likely to offer a high rate of
return. On distributive or equity grounds, virtually all of the benefits of
coconut research accrue to developing country producers and consumers.
Over half of the benefits accrue directly to producers in developing countries,
of which approximately 95% are smallholders, and the remaining benefits
accrue to the consumers. Thus, the major beneficiaries of coconut research
are the millions of smallholder coconut growers.

Another consideration is that of income security for smallholders. The
present price fluctuations are partially a result of erratic supply. Research
should aim to develop technologies to stabilize as well as to increase production. This includes identifying varieties able to remain productive under poor
environmental conditions, such as drought. This research would help to
stabilize the income level for coconut producers, and reduce poverty.

Potential Returns from Research

Over the past 25 years, coconut production has increased by about 2% per
annum as a result of new plantings. The average world yield of coconut has
shown no improvement over same period. In contrast, oilpalm and
rapeseed production have been increasing by around 10% per year, and
soyabean production by 5% per year (World Bank, 1991). These increases

are the result not only of increased areas but also of higher yield from new varieties.

In 1986, coconut provided 6% of the total world vegetable oil market, and in 1987 provided 5%. It is estimated that by 2000 it will provide only 4% of the market (World Bank, 1991). Thus coconut is steadily losing ground to other vegetable oils. Given the importance of coconut as a food and cash crop in many countries, the larger number of poor people involved in its cultivation and processing, and the lack of alternative crops able to fill the same ecological and social niche, it is important that efforts be made to make coconut a more competitive and a more economically attractive crop for smallholders.

The relative performance of coconut and oilpalm in recent years merits consideration. A concentrated research effort on oilpalm by both public and private sector agencies over the past 40 years has resulted in higher yielding clones which are able to be multiplied rapidly by newly developed micro-propagation techniques, and for which efficient production and processing technology is available. Oilpalm is thus competitive in the world vegetable oil market with the annual oilseed crops grown in temperate zones, and is well positioned to take advantage of the increasing world demand for fats and oils. The application of new technologies to oil palm cultivation has been facilitated by the strong participation of private sector plantation interests.

Coconut is predominantly a smallholder crop. In terms of the potential application of new biotechnologies, it is an 'orphan commodity' (Persley, 1990c). An effective research effort will require substantial public sector investments. The application of research results will also require the commitment of producing countries to provide the necessary technical services and appropriate policies to encourage smallholders to adopt suitable new technologies. These new technologies need to be designed specifically to meet the needs of smallholders.

Coconut breeding over the past 30 years has demonstrated that hybrids are capable of yielding copra up to 6.5 tonnes/ha/year between the ages of 10 and 20 years, under favourable environmental conditions and good management. (This contrasts with the world average yield of 0.5 tonnes/ha/year.) However, in some instances the acceptance of hybrids by smallholders has been subject to a number of real or perceived constraints and poor performance has been reported under some local conditions. These constraints to hybrid production and use, which relate to lack of adaptation to specific environments and difference in nut size, need to be further investigated.

The World Bank (1991) in assessing the current status of coconut research has identified seven areas where further research may be expected to yield high rates of return. These are:

1. Genetic improvement to develop a range of new varieties able to perform well in less favoured environments and under conditions of low management inputs. This will require expanding the genetic base of existing germplasm collections. New developments in modern biotechnology may contribute to improved breeding techniques, but foreseeable improvements will depend primarily on conventional breeding.

2. Micropropagation (tissue culture) techniques to enable elite palms to be cloned. Conventional methods of coconut propagation are slow and expensive, and only a limited amount of improved material is available.

3. Nutrient levels to determine the nutritional requirements of new varieties, and viable economic practices; for instance potassium and chloride applications, to enable smallholders to provide improved nutrition for their trees.

4. Multicropping studies to establish profitable and sustainable coconut-based farming systems. Although the coconut palm by its stature and growth habit is well adapted for use in multicrop farming systems, further study on the crop interactions in intercropping is required to make the best possible use of available land.

5. Pest and disease control Coconut is susceptible to many pests and diseases which restrict, or even preclude, its cultivation. Several lethal diseases cause the death of millions of palms each year, and for some, their causal organisms have yet to be identified. Others have been long known, but effective control measures that are based on integrated and environmentally acceptable techniques which minimize the use of pesticides still need to be devised.

6. Harvesting and processing to develop more economic methods for the production of copra and coconut oil. Traditional methods for copra productions are labour intensive, unpopular and increasingly expensive. High yields will require more small-scale mechanization, to enable the crop to be harvested and processed more economically.

7. New sustainable uses for the coconut palm (including the timber) need to be developed to enable the maximum profit to be derived by the full use of the crop (for both food and non-food uses), and increase returns to the farmer.

Rationale for International Support

The rationale for providing international support for research on coconut is based on:

1. The **growing demand** for vegetable oils and fats, especially in developing countries.

2. The **declining competitiveness** of coconut which is not able to take advantage of the expanding market for vegetable oils and is losing ground to other crops.

3. The low **average yield** of 0.5 tonnes/ha/year (copra equivalent) has not increased for the past 25 years. Production increases (of about 2% per annum) have come from the planting of new land. There are approximately

11.6 million ha of coconut land, spread over some 86 countries.

4. Coconut is a **smallholder crop**, produced largely from domestic consumption. Approximately 96% comes from smallholdings of 0.5–4.0 ha. There are more than 10 million farm families (about 50 million people) directly involved in its cultivation. A further 30 million people in Asia alone are directly dependent on coconut and its processing for their livelihood.

5. On a global basis, approximately 70% of the total crop is **consumed domestically**, and the balance exported. Many small countries with a low population export at least half their crop. For many island countries it is their only significant earner of foreign exchange.

6. The **major producing areas are in Asia and the Pacific** where about 85% of the crop is grown. There are 13 producing countries in Asia, and 19 in the South Pacific. Coconut is also locally important in the Americas where there are 33 producing countries. It is also important as a food and oil crop in the coastal areas of 22 countries in West and East Africa and the islands of the Indian Ocean.

7. Almost all the **benefits from coconut research** accrue to developing countries. Furthermore, the majority of these benefits go to the smallholder producers. The balance go to consumers in developing countries.

8. Research results in recent years suggest that there are areas from which there could be a **high rate of return on research investment**, analogous to that achieved from oilpalm. Appropriate methods will be required for the design of new technologies suitable for smallholders, if these returns are to be realized.

9. Coconut research is presently under-funded. There are several national research programmes, but, with few exceptions, they are not well-supported financially nor do they have sufficient appropriately-trained staff and facilities. Most suffer from a lack of continuity in funding, both from national sources, and from external agencies. Many small countries are not able to support a coconut research programme at all. At present there is no means by which small countries can access new technologies, or potentially higher yielding planting material for evaluation. Yet they could be active participants in an international programme. The present research effort is not addressing the needs of the crop internationally, nor capitalizing on the promising results from breeding and other areas of research, for the benefit of smallholders.

10. The **long-term nature of coconut research**, the history of discontinuity and lack of support in its funding, the prospects of high returns from research investments, and the likely distribution of research benefits to smallholder producers and consumers in developing countries, make coconut a suitable target for an international research initiative.

There is a need for a focused international research effort on coconut which would address the main problems facing the crop and make it more attractive to smallholders by giving them a higher return for their efforts. A number of high priority research activities can be addressed realistically only through an international programme, since they are beyond the scope of any individual national programme.

Any international initiatives should seek to be additional and complementary to existing national, regional and international efforts, and be perceived neither as a replacement, nor as competitive with them. The existence of an international effort would help to bring together a critical mass of expertise, resources and research capacity focussed on the needs of the commodity. It would also stimulate greater local investments in national coconut research programmes, which are presently underfunded, in relation to the importance of the crop to the economies of many countries. National programmes will also need to be supported through continuing bilateral activities. An international initiative would also provide training opportunities for national programme staff. The fragmentary nature of existing research, often resulting from a discontinuity of funding, would be addressed by an international initiative.

Research is required to increase the productivity of coconut trees by the use of locally adapted, high-yielding varieties, pest and disease management, and improved crop husbandry. Much of the research on the selection of locally adapted genotypes and appropriate agronomic practices is site-specific and should be the responsibility of national programmes. However, there are certain areas of production research which need to be undertaken at an international level, because they are beyond the scope of any one country but they could generate information of relevance to many countries.

The priority research areas which need to be addressed on a global basis lie within the areas of:

- genetic resources: collecting, conservation, breeding, and evaluation;
- disease and pest control;
- sustainability of coconut-based agroforestry;
- postharvest handling;
- socioeconomic research.

The problems appropriate for international support within each of these areas are discussed in the following sections.

Genetic Resources

Overview

Germplasm improvement consists of the collection and evaluation of genetic resources, and their utilization through selection and breeding. There is much opportunity for genetic improvement of coconut. The dominant, tall type of palm is heterozygous, with a great deal of unexploited variability from which to select. There is a need for new sources of resistance to the major diseases and for methods of rapidly transferring resistance genes into cultivated varieties.

Already available coconut hybrids have been shown to yield at least double some of the best available local material under favourable conditions. A well-planned and internationally financed programme to distribute potentially higher yielding material to many producing countries for evaluation under local conditions should offer high returns on research investment. There have been various international and regional activities among coconut breeders dating back to 1968. These activities and their outputs relevant to any new international activity are summarized in Box 5A, p. 113.

The long-term nature of breeding, and the need for large areas for field-based germplasm collections means that few countries have been able to mount effective breeding programmes. The existing programmes suffer from lack of continuity in funding (and a consequent stop–start approach to breeding), and often poor maintenance of their collections. All are breeding for national priorities. None has the designated responsibility to collect germplasm for use throughout the world, nor to breed hybrids and make them freely available for international testing in many countries.

Germplasm conservation and utilization

The current status of coconut germplasm collections is summarized in Table 5.1. The main existing collections are located in Côte d'Ivoire, India, Indonesia, Jamaica, Malaysia, Philippines, Solomon Islands, Sri Lanka, Tanzania, Thailand and Vanuatu. The maintenance of existing collections is an expensive item, beyond the reach of most national budgets.

The number of accessions in most collections is too few and not fully representative of the available germplasm. The content of the existing collections is uneven, with certain types being over-represented. For instance, the Malayan Dwarf accounts for nearly 50%, and some specific talls, such as the West African and Rennell, for about 15% of all accessions (Table 5.1). Plant breeders have tended to concentrate on material of known value. In most cases the collections have been made for a specific purpose, in support of a breeding programme. The collection in Jamaica, for example, was established for breeding against lethal yellowing disease.

There is a need for further collecting of coconut genetic resources. Much indigenous germplasm is under threat from genetic erosion, especially in areas where hybrids are likely to be widely planted. It is also important to widen the genetic base of existing collections for future breeding programmes.

IBPGR has supported coconut germplasm collecting in India, Indonesia, Malaysia and the Philippines. Unfortunately, IBPGR so far has not been able to extend its support to the conservation and evaluation of the collected

Table 5.1. Current status of selected national coconut genetic resources collections.

Country	Number of accessions	Total number of palms	Mean number palms per accession
India			
Kasaragod	121	3173	27
Indonesia			
Mapanget	20	833	42
Pakuwon	32	1786	56
Bone-Bone	50	2476	50
Côte d'Ivoire			
Marc Delorme	60	12213	204
Jamaica	35	?	?
Coconut Industry Board			
The Philippines			
Zamboanga	83	9552	118
Laguna	64		9
Baybay	7		
Tiaong	6		30
Albay	33		
Tanzania			
Zanzibar	13	1197	13
Mafia	13	1194	13
Thailand			
Chumphon	16	5808	363
Vanuatu			
Saraoutou	31	4805	155
Named Types			
Malayan Dwarf	34	20168	593
West African Tall	10	3801	380
Rennell Tall	7	2583	369
All Types	460	43037	104

Source: Information compiled by H. Harries. Data from Indonesia, Côte d'Ivoire, Philippines, Thailand and Vanuatu was drawn from IBPGR/SEAP and FAO/UNDP working group reports (IBPGR, 1976; UNDP/FAO, 1988). Data from Tanzania and Jamaica is from annual research reports. In Jamaica, the original number was reduced by lethal yellowing disease, increased by selected multiplication, then decimated by a hurricane in 1988. At Albay in the Philippines, the 1986 typhoon destroyed about 50% of palms.

material, some of which is being progressively lost from collections which are not well maintained.

In the consultations which formed part of this study, it was considered highly desirable for the effective implementation of an international research initiative on coconut that an internationally-funded germplasm research programme be established. Ideally, this would support a germplasm research unit with sufficient land on which to establish a coconut germplasm collection. The collection would be held under international auspices with one primary site, and several subsites on different continents. The multisite, international collection would be linked closely with existing national collections, with components of existing collections forming the initial basis of an international collection.

The primary site of the international collection should be located in an area usually free of natural disasters (especially typhoons); where there are no major disease problems that would preclude the exchange of material; and in the Asia/Pacific region, thought to be the centre of origin of the species.

Collaboration between the proposed international germplasm research unit and existing national and regional collections by means of an international network of coconut genetic resources collections will be essential. All material held under international auspices needs to have continuity of funding to ensure that valuable material is not lost. Free exchange of material among all participating countries would be a key requirement for all material held under international auspices.

An IBPGR working group on coconut germplasm has described the international status of coconut germplasm and breeding. The reports of this group (IBPGR, 1976, 1981) and other reports listed in Box 5A provide a useful basis for detailed planning of germplasm research needs, including the identification of future needs for germplasm collecting and conservation. There is also a need for an assessment of coconut germplasm evaluation techniques, so that results are comparable between countries.

Facilities and techniques for the international exchange of breeding material (particularly via embryos) need to be identified so that material can be exchanged safely, with appropriate quarantine precautions, and disease indexing where necessary. This matter was addressed at an FAO/IBPGR technical meeting in Indonesia in October 1991. The participants agreed upon guidelines for the safe exchange of coconut germplasm (FAO/IBPGR, 1992). It was recommended that these guidelines be used in the international coconut genetic resources network, being established under the auspices of IBPGR during 1992.

Coconut breeding

Major progress on coconut breeding has been made by the production of hybrids, often produced by crossing varieties of widely different geographic origin. The best tall varieties rarely produce more than 2 tonnes/ha copra. The best hybrids, under favourable conditions, can produce 6–6.5 tonnes/ha (World Bank, 1991).

The interval between generations in coconut is 5–10 years, using presently available techniques. An international effort is required to provide the continuity of funding which has been lacking for coconut breeding, and to enable access to potentially useful material by the many countries who are unable to support a national breeding programme.

There are many hybrids presently available from existing breeding programmes which could be tested widely, in an international germplasm evaluation network. Locally adapted varieties are required which combine high yield with tolerance to pests and diseases, and with desirable agronomic characteristics. Varieties which produce well in a poor environment and with low inputs are required for most situations. The genetic base of most breeding programmes needs to be expanded.

The adoption of standardized techniques of evaluation applicable in many national breeding programmes would be valuable. A major problem at present is that results from different countries are not comparable. A protocol for providing material for evaluation in many countries needs to be devised and adopted. Interchange of material and advice on agreed methodology for its evaluation would be a major contribution from an international coconut improvement programme.

A proposal for inter-country variety trials has been prepared for APCC by the Philippines Coconut Authority. This could provide the basis for the planning of initial inter-country evaluation of promising material.

Molecular biology as an aid to coconut breeding

Since germplasm improvement in coconut is a long-term process, new techniques to hasten the process would be especially useful. Development of such techniques based on new developments in molecular and cell biology may require collaboration between scientists in producing countries and those in other advanced laboratories elsewhere. Sponsorship of such linkages via commissioned research is an appropriate role for an international initiative.

For example, the development of new techniques based on molecular biology to assist in the classification of coconut types, and the early screening of material in breeding programmes (including new hybrids) would be valuable. Similarly, the development of other laboratory-based biochemical techniques (such as isoenzyme markers) to allow the characterization of naturally occurring coconut types (and eliminate duplicates from collections) would be useful.

Transformation and regeneration systems for coconut

The transfer of genes between species requires the development of effective transformation systems for the delivery of the new genetic information. Little work has been done on palm transformations, partly through lack of resources, but mainly because the tissue culture systems for plant regeneration are still too unpredictable. The recent advent of reliable transformation systems for some other monocots will provide an impetus to apply similar systems to coconut palms (Jones, 1990; Persley, 1990c), especially if regeneration protocols become available (see below).

Genetic mapping

There is an almost total lack of knowledge of the genetics of any of the characters which it might be of interest to modify, e.g. yield, quality, drought resistance, wind resistance and disease resistance. Such studies are vital to future applications of recombinant DNA technology to coconut (Jones, 1990).

Molecular biology provides useful tools for genetic mapping in the form of DNA probes, for example the use of restriction fragment length polymorphisms (RFLPs). Developments in genetic mapping could greatly reduce the number of plants carried in large and expensive field trials. There will be an increasing number of applications as the knowledge of the genetics of the crop develops. An RFLP map of coconut palm could provide the basis for genetic mapping of a large number of phenotypic characters within 2–5 years, and contribute to the choice of parents in directed breeding programmes (Jones, 1990).

Support is required for a coordinated effort in the production of a set of DNA probes and restriction enzymes for RFLP mapping of the coconut palm. This could be regarded as precompetitive research and could be funded by interested parties in both the public and the private sectors. It is important that techniques developed are accessible both to the individual companies and public research institutes concerned with conventional coconut breeding. Support will also be required to help these organizations develop in-house expertise, so that they can apply the new technologies in their breeding programmes.

Biotechnological developments in competitive crops (such as rapeseed) may be a future threat to the producers of palm and coconut oil. In particular the appearance of a temperate oilseed crop producing lauric and other short/medium chain fatty acids would pose a direct threat to coconut oil and palm kernel oil, which command a price premium as lauric oils. Such research is in progress in several laboratories worldwide. Some success has been reported in the USA in changing the fatty acid composition of rapeseed, by increasing its content of lauric acid. These developments should be monitored by agencies investing in coconut research, and research supported to ensure that coconut remains a competitive crop.

Cell and tissue culture

An international effort is required on coconut tissue culture to bring a critical mass of funds and expertise to the problems. The research needs in coconut tissue culture were reviewed by Professor Cocking of the University of Nottingham in September 1989, as part of the consultations on an international initiative for coconut research. The main research areas reviewed by Professor Cocking relate to: (i) clonal propagation; (ii) embryo culture; and (iii) cryopreservation. His observations on the current status and future needs in each area are outlined below.

Clonal propagation

The present methods for the propagation of coconut are slow and expensive. Even when elite material is available, it takes several years to produce sufficient quantity for use in a planting programme. Clonal propagation would allow the rapid propagation of high-yielding palms from hybrid populations and the multiplication of disease-resistant types. It would also allow greater use to be made of elite individuals in breeding programmes. It would facilitate *in vitro* germplasm storage. Also, the future application of genetic engineering to coconut improvement is dependent on having perfected tissue culture techniques to regenerate plantlets from single cells.

Clonal propagation would allow the wide phenotypic variability of the heterozygous hybrids to be stabilized, thus producing recognizable palm cultivars with stable characteristics (Jones, 1990). When successful, this technology will allow major improvements in yield and quality to be achieved. Although the techniques are now widely available for oilpalm, commercial development is delayed by the occurrence of flowering abnormalities in some of the first commercial clones. The cause of these problems has now been identified and clonal propagation of oilpalm (combined with quality control) is proceeding in field trials. Comparable techniques are required for the clonal propagation of coconut palm.

Cell and tissue culture techniques need to be developed much further to allow large-scale, routine propagation of coconut palm. The successful development of vegetative propagation for oilpalm, and the recent report of vegetative propagation of date palm suggest that clonal propagation is also technically feasible for coconut palm.

Clonal propagation of tissue from mature palms has been reported for a few plants (e.g. from India, the Philippines, Unifeld Laboratories, and Wye College, UK) (World Bank, 1991). Only a limited number of plantlets had been produced until 1990. More recently, it appears that ORSTOM/IRHO scientists in Montpellier, France may have succeeded in developing a more reproducible method for clonal propagation of coconut palm (de Nuce, personal communication).

Embryo culture

This technique was first demonstrated by Dr de Guzman and her colleagues in the Philippines for the Makapuno coconut type. Embryo culture techniques for other varieties have been developed in a joint programme between ORSTOM/IRHO and IBPGR, and in several national programmes (Whiters, personal communication). Embryo culture is important for: (i) the transport and preservation of germplasm; (ii) international exchange of germplasm (when combined with appropriate disease indexing techniques); (iii) multiplication of Makapuno-like nuts, which have a 'jelly' endosperm; and (iv) multiplication of hybrid embryos in breeding programmes. A protocol for culturing and establishing embryos needs to be made widely available for which there is a high success rate in the establishment phase.

Cryopreservation

There is also a need for improved methods for the long-term storage of coconut germplasm. The present field-based collections are expensive to maintain, and susceptible to natural disasters, especially typhoons. The current research being sponsored by IBPGR, IRHO and ORSTOM on *in vitro* conservation methods such as cryopreservation of coconut is important, as it would much reduce the cost and the land required for the conservation of coconut genetic resources.

International cooperation

Tissue culture is an area where a subject-specific research network could be established among interested scientists. An international initiative which had a significant contractual research budget could commission research with existing laboratories, as well as bring some new expertise into the network, especially in regard to clonal propagation. Although not all countries may need to be involved in the research directly, the results would be available and of benefit to all producing countries. The willingness of the private sector (where much of the current research is being done) to participate in this effort needs to be explored further.

Table 5.2. Distribution and severity of coconut diseases.

Disease	Asia				Pacific Islands	Americas	Africa
	India	Indonesia	Philippines	Sri Lanka			
Phytopthora	+	+ + +	+	+	+	+	+ +
Virus/viroids	−	−	+ + +	−	+ +	−	−
MLO	+ + +		−	−	−	+ + +	+ + +
Unknown aetiology*	+	+[a]	+[b]	+[c]		+ +[d]	+ + +[e]
Aspergillus (Aflatoxins)	+ +	+ +	+ +	+ +	+ +	+ +	+ +

Severity: + + + high, + + medium, + low; MLO, mycoplasma-like organism
Diseases of unknown aetielogy: [a]Natura wilt in Indonesia; [b]Socorro wilt in the Philippines; [c]Leaf scorch in Sri Lanka; [d]Hart rot in Latin America; [e]Cape St Paul wilt, Africa.

Disease and Pest Control

Disease distribution and importance

There are several lethal diseases which cause substantial losses of coconut palms. For example, it has been estimated that cadang-cadang disease has killed over 30 million trees in the Philippines since the disease was first recognized in the early part of this century. Annual loss in the Philippines has been estimated at US$16m (Hanold and Randles, 1991a; World Bank, 1991).

The distribution and relative importance of diseases of coconut are listed in Table 5.2. The diseases requiring further investigation are:

- lethal diseases (primarily mycoplasma-like diseases);
- virus/viroid diseases;
- *Phytophthora palmivora* (important in all coconut-growing regions).

The existence of several diseases whose aetiology remains unknown despite many years of research suggests the need for a concerted international effort, to facilitate collaborative research on the target diseases. Priority should be given to the lethal diseases, particularly those of unknown aetiology and to the virus/viroid diseases. The lethal diseases are killing millions of existing trees and precluding new plantings in some areas. The impact of viroids recently found to be widespread in some regions has yet to be assessed.

For example, the recent reports of a cadang-cadang-like viroid from several islands in the southwest Pacific and elsewhere suggests that the geographic distribution of coconut diseases is not well known (Hanold and Randles, 1991b). Surveys are required to clarify the present distribution of diseases. This is important for international quarantine and for the safe movement of germplasm, as well as for the development of national disease control programmes.

Lethal diseases caused by mycoplasma-like organisms

The lethal diseases caused by mycoplasma-like organisms (MLOs) pose the greatest threat to coconut in the future, since there are no economic control measures. Large areas of existing palms are at risk, and the threat of disease is a major disincentive to new coconut plantings in many parts of the world. Considering their similarity, worldwide distribution, and severe losses, a case exists for concerted international research on the lethal diseases of coconut.

The mycoplasma-like organisms implicated in these diseases are usually so difficut to detect, to culture *in vitro* and to transmit by artificial means that screening for resistance becomes difficult. Resistance is measured by field survival under conditions of high disease pressure. Research to date has been aimed primarily at breaking the disease cycle through identifying and manipulating disease vectors, or removing alternative hosts (FAO, 1991).

The spread of lethal yellowing disease, caused by a mycoplasma-like organism continues. The disease is associated with an insect-transmitted MLO. It has spread from Jamaica, to Florida and to Mexico. While Malayan Dwarf material is largely resistant in Jamaica, this situation may be changing since incidence has been reported to be unusually high in this variety in Jamaica and Florida. This necessitates continuing research to expand the disease resistance base for future breeding programmes.

The lethal disease in East Africa (Tanzania) has many similarities with lethal yellowing, and is presumed to be caused by MLO. Socorro wilt of the Philippines shows some similarities to lethal yellowing but its cause is so far unknown (FAO, 1991). The Kerala root (or wilt) disease in India has been shown recently to be associated with an MLO. The severe and rapidly-fatal Natuna wilt appears to be confined to a restricted area in Indonesia. Its causal agent also has not been identified.

Virus and viroid diseases

Cadang-cadang disease

This disease was first reported from the Philippines in the early 1930s, where at least 30 million trees have been killed. A related disease, 'tinangaja' in Guam destroyed most of the island's coconut palms. Both diseases have been shown to be caused by related viroids (Hanold and Randles, 1991a,b). The vector for the viroids is not yet known.

Other viroid diseases

Extensive field surveys of coconut palms, and oilpalms in the southwest
Pacific between 1987 and 1990 have demonstrated the presence of small
nucleic acids (viroids) similar to that which causes cadang-cadang in both
coconut palm and oilpalm at several locations (Hanold and Randles, 1991b).
The new viroid has been suggested to be associated with a symptom
previously known as 'genetic orange spotting' in oilpalm. 'Genetic orange
spotting' of oilpalm has also been reported from South America and West
Africa (Robertson *et al.*, 1968). The symptoms on coconut associated with
the presence of the viroids in the Pacific Islands were not nearly as severe as
those of cadang-cadang disease in the Philippines. However, some infected
palms were chlorotic, stunted, and had reduced yield.

Virus disease

Foliar decay, a damaging disease on introduced material in Vanuatu, has
been shown to be caused by a circular, single-stranded DNA virus (Rohde *et
al.*, 1990). The local Vanuatu Tall and the Vanuatu Red Dwarf are tolerant
of the disease.

Virus/viroid indexing techniques

Rapid detection methods based on the use of a nucleic acid probe have been
developed for cadang-cadang and related viroids, and for the virus causing
foliar decay in Vanuatu (Hanold and Randles, 1991a,b; Rohde *et al.*, 1990).
Rapid methods for screening for disease resistance to the major diseases,
particularly the lethal diseases caused by MLOs, are also required.

The results of the surveys in the Pacific Islands which showed a much
wider occurrence of viroids in coconut palms than had previously been
suspected, indicate that an indexing programme for propagating material of
coconut and oilpalm needs to be established, using molecular methods,
rather than symptoms, as a basis for testing for the presence of viroids and
viruses. Molecular-indexing methods for virus/viroid diseases will also be
essential to enable the safe international exchange of coconut germplasm.

Fungal diseases

Diseases associated with *Phytophthora* have increased in significance in
coconut during recent years. Bud rot and nut infection, caused by *P. palmivora* in southeast Asia, and by *P. heaveae* in West Africa, are causing
serious losses especially on hybrid and dwarf types. Research has shown that
chemical protection is possible and that there is potential for breeding for
resistance.

Table 5.3. Distribution and severity of selected coconut pests.

Pest	Asia				Pacific Islands	Americas	Africa
	India	Indonesia	Philippines	Sri Lanka			
Rhinoceros beetle	+ +	+ + +	+ + +	+	+ + +	−	+ +
Leaf-eating caterpillars	+ + +	+ + +	+ +	+ + +	+	+ + +	+ + +
Spike moth (*Thirathba* sp.)	−	+ +	+	−	+	−	−
Weevils (*Rhyncophorus*)	+	+ +	+ +	+ +	+ +	+ + +	+ +
Scale insects	+ +	+ +	+ +	+ +	+ +	+ +	+ +
Coconut mites[a]	−	−	−	−		+ + +	+ + +
Nematodes	−	−	−	−		+ + +	
Copra beetle	+ +	+ +	+ +	+ +	+ +	+ +	+ +

Severity: + + + high, + + medium, + low.
[a] *Acenia guerrenonis*

Nematodes

Red ring disease, caused by a nematode, *Rhadinaphelenchus*, transmitted by the palm weevil, *Rhyncophorus*, continues to kill both coconut and oilpalm in Latin America and the Caribbean (FAO, 1991).

Pests

The geographical distribution and relative importance of some selected coconut pests are shown in Table 5.3. Insects are the most serious pests, but a mite species and some mammals (rats, wild pigs, elephants) can also cause serious damage. The coconut is host to several hundred species of insect. Only about 1% cause sufficient economic damage to merit pest status; for these, however, considerable research is still required to develop environmentally safe and economically affordable control measures (Howard, 1991).

Insect pests may affect the leaves, resulting in defoliation; a specific part of the stem or bud, in which a single insect can cause the death of the palm; the roots; or the flowers and fruits. Other insects, not in themselves harmful, are vectors of lethal diseases.

In the past, old problems have been exacerbated and some new problems created by the injudicious use of broad-spectrum insecticides which have destroyed the natural balance between pest species and their natural parasites and predators. There are still chemicals being used against coconut pests which are being withdrawn from the market in industrial countries because they are no longer considered environmentally acceptable. The trend in pest control is towards integrated pest management, with the use of increasingly sophisticated methods of biological control. There is a need for strengthening of research on cultural practices which act against major pests both directly, and indirectly by benefiting natural enemies (Way, personal communication). As the factors associated with pest tolerance become better understood it is also likely that breeders will be able to develop coconut varieties which are more tolerant to pests (World Bank, 1991).

There has been considerable success in various countries in biological control of coconut rhinoceros beetles (*Oryctes* spp.) by baculovirus, especially where use of the virus is combined with improved hygiene in coconut plantings (e.g. removal of tree stumps where the beetles breed) (Way, personal communication). Methods of virus inoculum preparation, storage and simple techniques of inoculation have been devised for use by smallholders. Isolates of baculovirus have been found to differ significantly in their virulence and recent indications are that resistance by *Oryctes* to this valuable biocontrol organism may be developing. Some control of *Oryctes* has also been achieved through the fungus *Metarhizium* (FAO, 1991; Waterhouse and Norris, 1987).

Sustainability of Coconut-based Agroforestry Systems

Coconut-based systems are among the oldest farming systems in the world. Coconut is suited to multiple cropping, and is traditionally part of a long-term, multistorey farming system in many countries. The sustainability and continued productivity of these systems need to be addressed, since palms, including coconut palms, are under threat from excessive logging in various parts of the world. Numerous agronomic studies have been completed or are in progress in many countries. These studies have recently been reviewed by FAO (1991). The main findings of this review are summarized below (Turner, personal communication).

Extensive information, resulting from research, exists on coconut nutrition and the nutritional status of palms is readily monitored by foliar analysis. Information on optimal palm density in pure stand situation is also available for most ecological situations.

Most agronomic research on fertilizer experiments, spacing trials, and intercropping systems is site-specific. National coconut research programmes have undertaken much research in these areas over many years. These results are incorporated into national recommendations for coconut cultivation, and replanting and rehabilitation programmes. Techniques such as computer modelling and satellite imagery are helping to assess past experience in spacing trials, fertilizer trials and intercropping experiments, and make more efficient use of data from expensive and long-term research. Little research has been conducted on the physiological responses of coconut palm to environmental stress and competition. Such research would provide useful information for breeding programmes.

Experience of coconut-based multiple cropping systems has demon-strated advantages of intercropping with a wide range of additional crops, pastures and livestock such as bananas, cocoa, black peppers and pineapples (e.g. work in India, Sri Lanka, Thailand, Jamaica and elsewhere). Given the correct combination of crops and agronomic practices, coconut yield in intercropping systems has frequently been higher than in comparable stands of monocultured palms. Some of this additional yield has resulted from reduced weed competition, and soil fertility has been shown to improve through intercropping. While intercropping requires greater inputs of both labour and fertilizer, economic research has shown the benefits which have accrued in most instances. Much research continues into intercropping, particularly into the conditions necessary for sustainability of cropping systems. Similar beneficial results have been recorded from research on grazing by cattle, sheep and goats beneath coconuts.

Numerous studies have been conducted on nursery techniques and selec-tion to provide the best material for field planting, field planting techniques, and maintenance. Of particular value have been results from long-term fertilizer trials at numerous locations and establishment of a critical con-centration of the most important leaf nutrients, below which a response to applied fertilizers is likely to occur. This has been accompanied by the ability to relate responses not only to economic and environmental factors, but also to evaluate the beneficial or deleterious effects of agronomic practices on nutritional status. The amount of nutrients removed through cropping has been evaluated, especially for different hybrids. Research by the IRHO in particular has demonstrated the importance of chlorine in minimizing the adverse effects of water deficit in coconut plantings far from the sea. This research is on an inter-country basis in connection with selection for drought tolerance (FAO, 1991).

Many fertilizer trials have been conducted (e.g. in India, Indonesia, the Philippines, Thailand and Sri Lanka). Various nutrient deficiencies have been identified, often localized, with a requirement for fertilizer nitrogen being the one most widely recorded. What is now required is to use all existing informa-tion in an attempt to establish response curves of major nutrients under different environmental conditions, allowing more precise evaluation and use of leaf analysis results.

Nutrition research has shown the necessity for restoring soil fertility over vast areas from which harvested coconuts have removed nutrients for 60 years or more, with inadequate nutrient replacement. Such research has also furthered investigations into methods of rehabilitating old plantings, and choice of cover crops in new plantings. In the Philippines, a simulation model has been developed, which can be used to compare different coconut replanting strategies, including such aspects as planting material, spacing, and intercropping practices (FAO, 1991).

Agronomic research has enabled coconut to be cultivated successfully under certain conditions which were considered unsuitable previously. Suitable management and agronomic practices, derived from research findings, have resulted in increased yield from coconuts grown on peat swamp soils (e.g. in tidal zones in Indonesia). Such research opens up a considerable area for further coconut expansion, especially on coastal wetlands.

With the adverse effect of water stress on yield having long been established, allowance for this is made in research into yield forecasting. Irrigation has been shown to result in considerable yield increase, with very acceptable cost–benefit ratios, and detailed systems applicable to smallholders have been described. Moisture loss under rainfed conditions can be minimized during dry periods through use of cover crops and other herbage, mulches (e.g. coir, husk) and organic fertilizer. Several parameters have been measured and related to actual yield as a method of screening a range of cultivars for tolerance to drought.

Significant and rapid improvement in the production of many existing coconut smallholdings should be possible if economic means of enhancing the nutrition of the palms can be developed. The research priorities for improved nutrition of coconut palm are:

1. Establishment of response curves of major nutrients, particularly potassium, under differing conditions of water availability, to permit better interpretation of leaf analysis results.
2. Effect of fertilizer on drought tolerance, with emphasis on the effect of chlorine on stomatal activity.
3. Nutrient recycling and the benefits to be derived from minimizing the removal of plant products from the farm.
4. Effect of leguminous shrubs and trees in the interline to enhance soil conditions and nitrogen status with the production of firewood as an additional potential benefit.

Postharvest Processing

Overview

During the present study, Dr David Adair of ODA reviewed the current status and future needs for postharvest processing. His views are summarized below.

The majority of coconuts are consumed domestically, but a substantial volume (about 30%) enters world markets in the form of products derived by processing in the producing countries. The broad objectives of coconut research are to increase nut production to satisfy increasing domestic requirements and industrial requirements; to increase the size of the coconut processing industry and strengthen its competitive position; and to increase the size of the markets for the products. These objectives are interlinked: through research on germplasm improvement, disease and pest control, and agronomy, an increase in nut production is sought. Reduction of waste after harvest complements this approach. In many countries it offers substantial scope for effective increase in domestic supply of food/animal feed, and of the marketable surplus for use as raw material by processing industries. Advantage cannot be taken of that surplus unless expanded markets can be secured for the products of the industries.

There are two main foci for international postharvest research efforts: (i) health of coconut consumers worldwide; and (ii) value added in coconut processing. The principal health concern at present is that badly handled and badly processed coconut foods and feeds contain toxic fungal metabolites such as aflatoxins. The value-added component will be increased by expanding the market for coconut products overall and by increasing the proportion of nuts which is converted into the higher-priced products, such as desiccated coconut.

There is undoubtedly scope for greatly increased bilateral activities, dealing with problems which are highly country-specific. However, there are many problems which affect coconut producers at large. For example, the extent of the aflotoxin problem has yet to be elucidated, and effective strategies for dealing with it defined. A reliable field instrument for rapid determination of the moisture content of copra is not available at present. Improved methods for rapid detection of *Salmonella* in coconut products are required. There is a widespread need for more efficient, less laborious, methods for domestic processing of coconut in rural areas. Even countries which have a long tradition of skilled and highly efficient manual dehusking now anticipate a need for mechanical dehuskers in the near future.

The position of coconut products relative to their competitors needs to be monitored, and addressed continuously through research and the dissemination of research findings. Research is also necessary to provide a foundation for the defence of coconut products against attacks made on the basis of speculative interpretations of scientific, technical, or nutritional data. Equally, research is necessary as a basis for reliable advice on claims made about innovations in processing technology, and to generate more reliable information about coconut oil in the human diet (see Box 2B, p. 43).

The present practice of producing copra, then coconut oil, is wasteful. However, this should not be regarded as justification for abandoning copra production in favour of alternative technologies, because currently the infrastructure of copra importers is geared to handle this product. The need is to improve the efficiency of copra production worldwide. This must be the main objective of the coconut industry in the short and medium term, and work towards it should address copra production and oil expression from copra as an integral system. This is the only system available at present which offers market outlets for kernel products commensurate with the volume of nuts available for processing. There is a widely held view that improvement in copra production requires only the application of existing knowledge and technology. This is mistaken; there is a major element of research in the work required.

Small-scale processing techniques which allowed the production of coconut oil (and other products) at the village level would also be useful, especially in isolated areas with infrequent shipping services. The development of simple, mobile dehusking devices is also a priority. These issues are important for smallholders, who require less labour-intensive means for processing coconut. Etherington (1988) has reviewed the possibilities for research into new processing technologies which would maximize returns to smallholders. Hagen (1991) has also described the energy systems required for small-scale coconut processing, and recommended a sustainable coconut energy system appropriate for small island nations.

In the medium to long term, expansion of markets for kernel products, such as desiccated coconut and coconut cream, which confer greater added value must be sought. Here a marketing and processing research interface with the food processing industry in importing countries is needed.

With good prospects for expansion of markets for activated carbon, research is required to optimize the use of the coconut shell byproduct, which is often used wastefully in traditional industry. Husk is also an under-utilized resource. There is a need for research on both processing of coconut fibre products and the requirements of the fibre market.

The value of international cooperation in research on coconut wood utilization has been demonstrated by a number of programmes, notably the work involving the Philippines, and Tonga, with support from New Zealand and FAO. There is a need for continuation and expansion of this type of cooperation, linked to the removal of old trees in replanting programmes.

Research priorities

The postharvest research requirements for coconut are summarized below. The present research needs concern the domestic processing of coconut, and the more efficient production of copra, coconut oil, desiccated coconut, coconut cream, shell, fibre products and wood. Emphasis in any international initiative should be given to the processing of the major products (copra and coconut oil), since improvements here will be relevant to producers in many countries. The specific research objectives are:

1. **Domestic processing of coconuts**. More efficient, less wasteful, and less

laborious methods of extraction of cream and oil from fresh kernel.

2. Copra production. Reliable sampling methodology; reliable and rapid methods of assessing moisture content of kernel, suitable for field use. Standardized methods of aflatoxin determination. Improved equipment and procedures for copra making, with increased energy efficiency. Processes for detoxifying aflatoxin-contaminated copra, unrefined coconut oil and oilcake.

3. Coconut oil. Improved understanding of the role of coconut oil in human nutrition; additional industrial uses of coconut oil.

4. Desiccated coconut. Better market information as a guide to technological R&D requirements; clearer product standards; improved hygiene; better energy efficiency in processing; improved methods of *Salmonella* detection.

5. Coconut cream. Better market information as a guide to creation of new outlets; cleaner product standards; development of diverse products using coconut milk or cream as the raw material.

6. Shell. Integration of shell charcoal production with kernel-processing operations to optimize use of the shell's energy and to reduce atmospheric pollution.

7. Fibre products. Better market information as a guide to technological research needs; optimization of mechanical dehusking; improvement of safety and efficiency in extraction of fibre from husks; adaptation of fibre processing equipment to new scales of operation to suit potential new users; reduction of negative effects on the environment through identification of viable technology for the use of coir dust.

8. Stem. Better market intelligence as a guide to the expansion of stem utilization; better technology for the extraction, preservation, and utilization of stem fractions.

Socioeconomics

In the consultations for this study, many agreed that there was a socio-economic research component required in the areas of germplasm improvement, pest and disease control, the sustainability of coconut-based farming systems, and postharvest processing. This is to ensure that the real socioeconomic constraints of smallholders are being addressed, and that new technologies being developed are suitable for smallholders to adopt.

Specific topics for socioeconomic research have been identified as:

1. suitable multicropping systems for traditional varieties and hybrids;
2. the economics of smallholders using inputs, such as fertilizers;
3. the social and economic constraints to participation in replanting and/or rehabilitation programmes, also need to be addressed. Ways need to be found to maintain income of farmers while the new trees are coming into bearing (perhaps by phasing the removal of old palms). Even for the early-bearing hybrids, the first nuts are not harvested for about 4 years.

Priority Areas for International Support

The final series of consultations in this study focussed on identifying, from the above research needs, the priority areas suitable for international support. These are areas where the research would be international in character, i.e. it addresses problems of importance to many producing countries, but which are beyond the financial and/or scientific scope of any one country to solve. Based on these criteria, the targets and the priority areas for international support are shown in Box 5B, p. 117.

CGIAR Priorities and Future Strategies

Consideration of coconut research by the CGIAR commenced in the mid-1980s, when TAC in its then review of CGIAR priorities and future strategies, identified coconut as one of three new priority commodities for support through international research (the other two being vegetables and fisheries). In its 1986 report to the CGIAR, TAC considered that the establishment of an international coconut research network would provide a focal point for collaboration and for donor support. The value of coconut as an ecologically sound food and cash crop suitable for smallholder cultivation, the geographical diversity of its production, the potential for further research, the uncoordinated and underfunded research effort to date, and the need to solve important disease problems were strong reasons for coordinating and strengthening current research efforts. TAC therefore, encouraged the creation of a research network to strengthen and coordinate coconut research and supported CG system involvement in such a network (CGIAR, 1986). The CGIAR accepted this recommendation at its meeting in Ottawa in May 1986, and requested TAC to explore further the possibility of establishing a CGIAR-supported initiative on coconut, and the form such an initiative might take.

TAC's subsequent consideration of coconut research was included within TAC's broader review of expanding the scope of the CGIAR system to include forestry and agroforestry, as well as several other existing international agricultural research centres. Between 1988 and 1990, TAC considered three reports prepared by ACIAR which addressed the need and feasibility of establishing an international research initiative on coconut. The content of these reports has been summarized in this and the preceding chapters.

After consideration of all the issues, TAC recommended to the CGIAR in 1990 that coconut be included within the CGIAR portfolio of activities. TAC further recommended that the research areas which warranted an international effort were in the fields of:

1. germplasm collecting, conservation, evaluation and enhancement;

2. the control of diseases and pests, especially the lethal diseases;

3. the productivity and sustainability of coconut-based agroforestry systems;

4. the need for greater efficiency and added value in postharvest handling and utilization;

5. socioeconomic issues, especially the factors which influence farmers' participation in rehabilitation and replanting.

With regard to the possible institutional arrangements, TAC saw a clear need for networks with a strong 'enabling' component to fund research. TAC considered that provision for this enabling component should be made through close cooperation with existing development agencies. Further, TAC stressed that close cooperation with IBPGR would be required with respect to coconut germplasm collecting and conservation. TAC also recommended that IBPGR be invited to establish and manage a small germplasm research unit for coconut in the Asia/Pacific region (CGIAR, 1990). The text of TAC's deliberations and its recommendations on future international support for coconut research are given in Box 5C, p. 119.

In the current report on the CGIAR priorities and strategies, TAC emphasized that its earlier recommendation to include coconut in the CGIAR portfolio was based on its importance as a smallholder multipurpose tree crop in several farming systems throughout the tropics. TAC noted the prospects of high returns from research investments, the benefits to low-income producers, and the lack of continuity in historical research efforts. Coconut was also an important crop for the sustainability of agricultural production in coastal ecosystems. TAC reiterated its recommendation for the inclusion of coconut within the CGIAR portfolio of commodities (CGIAR, 1992). The recommendations of this report were adopted by the CGIAR at its meeting in Istanbul in May 1992.

The CGIAR has agreed to the inclusion of coconut within its portfolio of activities, and to the five priority areas appropriate for international support, as recommended by TAC. It also agreed that genetic resources should be given first priority, and encouraged IBPGR to strengthen its work on coconut genetic resources.

In regard to the means of implementation of a broader international initiative on coconut research, which addressed the five priority areas identified by TAC, the CGIAR recommended that the question of institutional mechanisms should be taken up within the context of the CGIAR's expanding support for agroforestry and forestry, and the inclusion of ICRAF and the newly established Centre for International Forestry Research (CIFOR) in the CGIAR system.

The institutional options for establishing an international initiative on coconut research, either within or outside the CGIAR system, are discussed in the following chapters.

BOX 5A: Selected International and Regional Activities on Coconut Germplasm 1968–1990.

Year	Activity	Lead Agency
1968	Introduction and exchange of coconut germplasm 1959–1966	FAO Technical Working Party on Coconut Production, Protection, and Processing

This report gives, in tabular form and in chronological order, information from many sources on introduction and exchange of coconut germplasm effected from 1959 to 1966. Much of this period coincided with the lifespan of an FAO Regional Coconut Improvement Project and followed the First Session of the FAO Technical Working Party on Coconut Production, Protection and Processing, held in Trivandrum, Kerala State, India, in November–December 1961.

Year	Activity	Lead Agency
1969–1981	Coconut Breeders' Consultative Committee	FAO Technical Working Party on Coconut Production, Protection, and Processing

The Coconut Breeders' Consultative Committee was set up in 1968 as an *ad hoc* subcommittee of the FAO Technical Commission on Coconut Production, Protection, and Processing. Three technical working party meetings were held (India, 1961; Sri Lanka, 1964; Indonesia, 1968) before the Breeders' Committee was established, and two afterwards (Jamaica, 1975; Philippines, 1979). The FAO Technical Working Party on Coconut Production, Protection, and Processing was abolished in 1981.

Following the format used in the 1968 report, tabulated data on coconut germplasm introduction and exchange, together with written information on evaluation and utilization, were compiled in ten issues of a Yearly Progress Report on Coconut Breeding between 1969 and 1978. This became the 'yearly progress report on coconut research and development', for three issues from 1979 to 1981 and continued to include the contributed germplasm information. At its zenith a 40-page report was distributed to 224 individuals or organizations in 54 different countries.

Year	Activity	Lead Agency
1979	Multilocational testing of promising hybrids	UNDP/FAO

A detailed protocol for multilocational testing of F_1 hybrid coconuts was prepared. Countries in Asia and the Pacific expressed interest in the proposal but it did not proceed.

Year	Activity	Lead Agency
1981	IBPGR Directory of Germplasm Collections: Industrial Crops (Coconut)	International Board for Plant Genetic Resources (IBPGR)

Box 5A: Continued

This directory was produced towards the beginning of the IBPGR's active interest in coconuts when it assisted with germplasm collection by Indonesia, Philippines, Malaysia and Mexico between 1980 and 1983.

Year	Activity	Lead Agency
1987	IBPGR/SEAP Working Group on Palms and UNDP/FAO Working Group on Genetic Improvement	IBPGR Cooperative Regional Programme for Southeast Asia and FAO/UNDP Project RAS/UNDP Project RAS/80/032 Coconut Production in Asia and the Pacific

A joint meeting was held at which genetic resources activities in India, Indonesia, Malaysia, Papua New Guinea, Philippines, Sri Lanka and Thailand were reported. Recommendations were made on collection and characterization, evaluation, multiplication, conservation and documentation, germplasm exchange and utilization, research and regional joint projects.

Year	Activity	Lead Agency
1988	UNDP/FAO Working Group on Genetic Improvement	FAO/UNDP Project RAS/80/032 Improved Coconut Production in Asia and the Pacific

A second meeting of coconut breeders was held under the auspices of the FAO/UNDP regional coconut project, with participation from Indonesia, Malaysia, Papua New Guinea, Philippines, Sri Lanka, Thailand, Vanuatu and Western Samoa. FAO and IRHO were also represented.

The objectives of the meeting were to establish a protocol for varietal trials, discuss and recommend appropriate quarantine procedures and possibilities for expanding the use of embryo culture to facilitate the safe exchange of plant material; review recent germplasm exploration, identify areas for collecting and propose a programme for phased exploration and collection; and discuss accessibility of collections held in the region. The FAO/UNDP regional coconut project has since ceased.

Year	Activity	Lead Agency
1989	International Registration Authority for Coconut	International Society for Horticultural Science (ISHS)

The ISHS interest in making a checklist of coconut cultivars was first expressed in 1973. In addition to clarifying the confusion over identifying named types of coconut there was the stated intention of registering the F_1 hybrids which were being produced in commercial quantities for the first time. The initiative was not followed up at that time. Renewed interest by ISHS in 1988 has been supported by the International Palm Society and a Registrar has been appointed. There are international registration authorities for many ornamental plants and for fruit tree crops, such as mango, but none for any palm genera.

Box 5A: Continued

Year	Activity	Lead Agency
1989	Coconut Germplasm Research Unit, Indonesia	UNDP/FAO

A proposal for a national germplasm collection in Indonesia was submitted to FAO, but did not proceed.

Year	Activity	Lead Agency
1989	Latin American Regional Coconut Germplasm Collection	UNDP/FAO

A proposal for research in Mexico into lethal yellowing disease was submitted to FAO. If accepted it would involve varietal testing for resistance. A regional germplasm collection in a disease-free country in the region would be required. Interest in the possible location of the regional collection has been expressed by Costa Rica.

Year	Activity	Lead Agency
1990	Inter-country Trial of Promising Coconut Hybrids and Cultivars	Asian and Pacific Coconut Community

A proposal for multilocational, inter-country trials of promising coconut hybrids and cultivars has been submitted to APCC by the Philippines. Similar to a project proposed to FAO in 1979 (see above) it proposes to take advantage of embryo culture techniques to promote safer and faster exchange of important coconut germplasm.

The immediate objectives are to undertake performance tests of promising coconut varieties, hybrids and cultivars, to identify widely adaptable coconut ecotypes for use in planting and replanting programmes and to study their reaction to pests and diseases in the region.

Year	Activity	Lead Agency
1990	Training course on coconut collection, conversation and evaluation	IBPGR Regional Office for South and Southeast Asia

The IBPGR supported a course in the Philippines with course tutors from IRHO and India. Participants came from several Asian countries.

Year	Activity	Lead Agency
1991	First African Coconut Seminar	BUROTROP

A meeting at Arusha, Tanzania produced recommendations for an African coconut genetic resources network, for multilocation variety trials, for disease

Box 5A: Continued

resistance and drought tolerance breeding programmes, and for an African coconut breeding working group.

Year	Activity	Lead Agency
1991	Safe movement of coconut germplasm	IBPGR & FAO

A meeting on the subject was held in Ciloto, Indonesia to produce a set of guidelines for the safe movement of coconut genetic resources.

Year	Activity	Lead Agency
1991	International Workshop on Coconut Genetic Resources	IBPGR

A meeting on the subject was held in Cipanas, Indonesia. The establishment of Coconut Genetic Resources Network was recommended and a ten member Steering Committee was formed (see Chapter 7).

Sources: Information collated by Dr Hugh Harries, from: FAO (1969–1981); IBPGR (1981); IBPGR/SEAP (1987); Pieris (1968); UNDP/FAO (1988)

Box 5B: International Coconut Research Priorities

1. Genetic Resources
To support the conservation, evaluation and utilization of coconut genetic resources by:

- collection and conservation of the genetic resources of the coconut palm
- techniques for the safe international exchange of germplasm (e.g. via embryo cultures, pollen)
- inter-country exchange and evaluation of germplasm using common methodologies
- new biotechnologies to shorten the breeding cycle and introduce useful new characteristics (e.g. transformation and regeneration systems; clonal progapation; genetic mapping)

2. Disease and Pest Control
To control the major lethal diseases by:

- identification of the causal agents of lethal diseases of unknown aetiology in Asia, Africa and the Americas
- improved diagnostic techniques for mycoplasma-like organisms
- indexing protocols for virus/viroid diseases to facilitate the safe international exchange of coconut germplasm
- assessment of the distribution and significance of non-lethal viroids in palms

To foster integrated pest management by:

- development of cultural practices which favour integrated pest management
- inter-country exchange of natural enemies against major pests
- surveys for natural enemies of the major pests in the Indo/Pacific areas
- identification of improved strains of viral and fungal pathogens for the biological control of the rhinoceros beetle

3. Sustainability of Coconut-Based Agroforestry Systems
To understand the principles of coconut-based ecosystems which govern the following:

- multistorey systems, to understand the principles of successful multi-cropping
- nutrient supply, through nutrient recycling
- light interception, through adjusting tree density to maximize light use by intercrops
- water relations, through interaction between coconut and its intercrops

4. Postharvest Processing
To improve 1) the efficiency and quality of copra and coconut oil production, and 2) increase the added value on coconut processing by:

Box 5B: Continued

- improved small-scale processing methods for copra and coconut oil production
- improved detection and detoxification methods for aflatoxins
- increased quality of value-added products such as desiccated coconut, coconut cream, shell and stem

5. Socioeconomics
  To identify the socioeconomic constraints affecting coconut producers by:

- identification of the social and economic constraints to participation by smallholders in replanting and/or rehabilitation programmes
- factors affecting adoption/non-adoption of new technologies
- suitable multi-cropping systems
- the economics of fertilizer and pesticide use

Box 5C: TAC's Recommendations to the CGIAR on International Support for Coconut Research

Background

The coconut palm is a pan-tropical crop, grown on approximately 11.6 million ha in 82 countries. Many of the producing countries are small islands in the Pacific, Caribbean and Indian Oceans. Coconut is both their primary subsistence crop, and their only significant source of export earnings. There are few, if any, alternative crops which can substitute for coconut in these countries. In many island countries it provides 50% or more of total export earnings.

Coconut is predominantly a smallholder crop, with at least 96% of total world production coming from smallholdings. About 70% of the total crop is consumed in the producing countries. The crop can be grown in harsh environments, such as atolls, high salinity, drought, or poor soils. It plays an important role in the sustainability of often fragile ecosystems in island and coastal communities. Coconut is used as a source of food, drink, fuel, stock feed and shelter for village communities. It is also a cash crop, able to be used to produce many items for sale, at either the local, national, or international level. The main internationally traded products are copra, coconut oil, copra meal, and desiccated coconut.

The 1986 TAC review of CGIAR Priorities and Strategies identified coconut as a priority commodity for international support. Subsequently, the CGIAR requested TAC to explore the desirability of establishing an international research initiative on coconut, and the form such an initiative might take. TAC then considered the current status and future trends for coconut within the context of the world fats and oils markets; the importance of coconut as a subsistence crop, as a cash crop, and as a component of long-term farming systems; current research programmes; future needs for coconut and opportunities for further research; possible options for an international research initiative; and possible institutional mechanism by which such an initiative might be implemented.

Relative priority of coconut

Coconut is a major crop in Papua New Guinea and the Pacific Islands, the Philippines, Indonesia and South Asia. It is locally important in coastal regions throughout the remainder of Asia, in West and East Africa, Mexico, Central America and the Caribbean Islands. In addition to the value of coconut production for domestic use and export earnings in those areas, there are other reasons why coconut would be appropriate for CGIAR support. These include its significant contribution to agricultural GDP and foreign exchange earnings of many small, mostly island countries; the important role coconut plays in intercropping systems, and the sustainability of coastal lands and islands; and the central role coconut plays in village life by providing many items for food, shelter and clothing.

Future needs and opportunities for coconut

There are four major constraints to increased coconut production in developing countries: the low productivity of many coconut trees due to their age and poor

Box 5C: Continued

nutrition; the failure of many replanting programmes designed to replace old trees which are beyond their productive life span (about 60 years); the fluctuating productivity due to variable environmental conditions; and the inefficient handling and processing with a low farm-gate price to smallholders. The needs are to increase the productivity of the crop by the use of locally adapted high-yielding, pest and disease tolerant varieties in any replanting or new planting schemes; to increase the productivity of existing plantations by encouraging better agronomic practices, including the control of diseases, insects and weeds, the use of fertilizers, and the identification of profitable and sustainable intercropping systems; to develop improved methods of handling and processing coconut; and to diversify the coconut products traded and actively promote new products in the marketplace, so as to utilize fully the potential of the crop.

The opportunities arise from the increasing demand for oils and fats and animal feed sources, particularly in developing countries as incomes rise, and the ability of the coconut tree to produce a wide variety of useful products.

Rationale for research

The increasing importance of coconut to meet the growing demand for vegetable oils and fats in producing countries; the continuing premium prices paid for the lauric acid oils (coconut and palm kernel oil), primarily for their industrial uses in soaps and detergents; and declining competitiveness of coconut, provide the rationale for increased research effort. Research is required to make coconut more competitive by increasing its productivity in a manner analogous to what has occurred with oilpalm. Virtually all the benefits of coconut research will accrue to developing country producers and consumers.

Coconut is the major tree-crop component in several agroforestry systems throughout the world. Its wide use in home gardens is probably not reflected in official statistics for area under cultivation, volume of production and total value of production. Research should aim at developing technologies to stabilize production and thus contribute to regular income levels for coconut producers, and the reduction of poverty.

Coconut breeding in several countries over the past 30 years has demonstrated that hybrids are capable of yielding up to 6 tonnes/ha/year of copra, under favorable conditions (cf. world average yield of 0.5 tonnes/ha/year). Progress has also been made in the identification of the causal agents of diseases of previously unknown aetiology, such as cadang-cadang disease in the Philippines and lethal yellowing disease in the Caribbean. Nutritional studies have shown that coconut responds to fertilizer application, particularly potassium and chloride. Intercropping and the use of cattle under trees has shown that the total productivity of the coconut lands can be improved, while still maintaining the long-term sustainability of the system.

These results suggest that a well-organized and adequately funded international research effort could yield high returns on the investments. The long-term nature of coconut research, the history of discontinuity and lack of

Box 5C: Continued

support in its funding, the prospects of high returns from research investments, and the likely benefits to smallholder producers, make coconut a particularly suitable target for an international research initiative.

Recommendations

TAC concluded that, on the basis of many of the criteria that relate directly to the revised CGIAR mission statement, the priority accorded to coconut should be higher than would be the case if the commodity were assessed simply on the basis of its estimated global worth in economic terms. TAC therefore recommended to the CGIAR that coconut be included in the CGIAR portfolio of activities. The priority research areas that warrant an international effort are:

1. germplasm collect on, conservation, evaluation and enhancement;
2. disease and pest control, especially in regard to the lethal diseases;
3. the productivity and sustainability of production systems, in which coconut is a major or minor component;
4. postharvest handling and utilization;
5. socioeconomics.

TAC considered that the most urgent need was for an international programme to create a living collection of coconut germplasm and make it available to national breeding programmes. This had been identified by the TAC panel as an area of high priority with Indonesia suggested as the preferred location for a living international collection.

TAC recommended that the CGIAR should support a small initiative on coconut research, and that IBPGR should be asked to strengthen its work on coconut genetic resources through a crop-specific network of national programmes. IBPGR should also be invited to establish and manage a coconut germplasm unit in the Asia/Pacific region.

TAC envisaged that the work of the germplasm unit would include collection and characterization of coconut germplasm, as well as the establishment and maintenance of a living collection. Its primary purpose would be to provide source material to national breeding programmes. It would conduct research in collaboration with other institutions on such topics as tissue culture, micropropagation and gene transfer.

Although there are many bilaterally and multilaterally funded coconut research and development projects, all currently operate for relatively short periods (usually 3–5 years), and are rarely linked with one another. The discontinuity of funding is not conducive to a sustained increase in coconut productivity. If an increase in the productivity of coconut is to be achieved, a critical mass of funds and research capacity over a sustained period will be needed.

TAC carefully examined the advantages and disadvantages of various institutional options by which an international research programme could be established. Given the number of small countries in which coconuts are important, TAC saw a clear need for networks with a strong enabling component. TAC considered that it would be essential for the enabling function to be kept

Box 5C: Continued

in mind and provision made for meeting it, preferably through close collaboration with an existing development agency. Further, TAC stressed that whatever institutional option was adopted, it should collaborate closely with IBPGR with respect to coconut germplasm collecting and conservation. In this regard, TAC recommended that IBPGR should prepare a detailed proposal in regard to the germplasm improvement activities, and that this should be presented to TAC as part of the institute's mid-term programme and budget.

Source: CGIAR (1990, 1992).

Chapter six:

Institutional Options for International Support

'The preferred option from the consultations was to form a
new body for coconut research.'

Institutional Options

During this study, four possible options for the provision of additional
support to address the problems facing coconut on a global basis were
considered. These were: (i) to provide additional bilateral support to national
research programmes; (ii) an to establish an international coconut research
centre; (iii) an international coconut research network; or (iv) an international coconut research council. The advantages and disadvantages of each
option are summarized below:

Option 1: Additional bilateral support for national programmes

The advantages of the provision of additional support to national programmes
are that it would: (i) build on existing research capacity and facilities in
national programmes; and (ii) allow decentralized activities, in major and
minor producing countries.

The disadvantages are that: (i) it does not bring a critical mass of
scientific expertise and resources to the global problems of the crop; (ii) it has
no international focal point; (iii) there is no global view of research needs;
and (iv) it has no in-house research capacity.

Option 2: International coconut research centre

An international coconut research centre could be established in the mode of
existing international commodity research centres such as the International
Rice Research Institute (IRRI), with centralized research facilities in a
coconut-producing country.

The advantages of this option are that it would: (i) provide a focus for research, training, and documentation on coconut; (ii) assemble a critical mass of scientists and resources at one location, under effective management and with continuity of funding; (iii) establish a centre of excellence for research; (iv) provide a global view of research needs; and (v) guarantee free availability of research results and improved germplasm for international exchange.

The disadvantages are: (i) its high cost, especially in the establishment phase; (ii) the fact that it does not build on nor support present national research capacity and facilities at existing research centres; and (iii) its geographic concentration of staff and resources: pests and diseases of coconut differ throughout the world, and cannot be brought to a central location to be studied.

Option 3: International coconut research network

An international coconut research network could be established among national research institutes and other interested regional and international bodies, to coordinate existing activities.

The advantages of this option are that it: (i) maximizes the use of existing programmes and institutions; (ii) provides a cost-effective mechanism; (iii) develops a strong sense of equal partnership amongst the network members; (iv) could provide a global view of research needs; and (iv) allows decentralized activities in major and minor producing countries.

The disadvantages are that: (i) it has been previously attempted in the Asia/Pacific region with little success; (ii) there would be no in-house research capacity and no scientific backstopping for the network; and (iii) continuity of funding would continue to be a problem since most additional research would still need to be funded from national and bilateral sources on a short-term, project basis.

Option 4: International coconut research council

A new body is proposed, termed here an 'International Coconut Research Council' to identify, support, promote, and undertake research on priority problems of international significance. It would be able to: (i) commission additional research on priority research problems of global significance, to be undertaken on a contractual basis by national programmes, regional organizations, or other advanced laboratories; (ii) undertake research itself on a limited scale, especially in relation to germplasm improvement; (iii) establish subject-specific research networks; and (iv) encourage the formation of regional networks.

The advantages of this option are that it: (i) builds on existing research capacity by providing additional funds to enable individual national programmes to undertake research of relevance to many countries; (ii) provides a mechanism to internationalize the activities of existing research institutions; (iii) brings additional scientific and financial resources and leadership to research; (iv) identifies globally important research priorities; (v) provides for continuity of funding; (vi) provides financial support for national and regional germplasm collections and breeding programmes of international significance; (vii) facilitates collaborative research amongst scientists in different countries; (viii) allows participation by both public and private sector organizations; (ix) allows participation by small countries with no national research programme in an international programme; (x) provides a decentralized approach; and (xi) has some in-house research capacity.

The disadvantages are that: (i) research leadership and management would be more difficult than in an international commodity centre at one location; and (ii) it will be more difficult to avoid a top-down perception from large to small countries and agencies.

Preferred Option

The establishment of an International Coconut Research Council (Option 4) was the preferred option in the consultations which formed part of this study. There was a widely-held view that the other options described above would not bring sufficient focus and resources on major problems (Option 1); were too centralized (Option 2); or too decentralized (Option 3) to be effective. In addition, Option 1 (additional support for national programmes) and Option 3 (an international network) had been attempted in the past, with little success.

The majority view was that if there was serious interest in sponsoring an international research initiative on coconut, a new institutional model needed to be designed. Considerable thought was given to the details of how an international research initiative could be established which would highlight the advantages of various other options, minimize their disadvantages, and bring some new modalities to the support of international agricultural research.

The ideal scenario for the proposed International Coconut Research Council is described in the following section.

International Coconut Research Council

Goals

The purpose of the proposed International Coconut Research Council would be to identify, support, promote and undertake research on priority problems of international significance. The new body would:

1. Conduct research itself on a limited scale, especially in relation to germplasm conservation and utilization.
2. Enable additional research to be undertaken on a commissioned basis by national programmes, regional or international organizations, or other advanced laboratories, within an agreed global programme of priority problems.
3. Organize subject-specific research networks amongst active research workers, on problems of international significance.
4. Establish regional networks, to identify the priority problems requiring additional research efforts, and to facilitate the distribution of research results to all coconut-producing countries.

Such an international initiative could:

1. Identify international research priorities for coconut, on problems relevant to many producing countries, which cannot be addressed adequately by any one country.
2. Build on existing research capacity by providing additional funds to enable national, regional and international programmes to undertake research of relevance to many countries.
3. Provide additional support for selected genetic resources collections and enable valuable material to be held under international auspices.
4. Provide additional support to breeding programmes of international significance.
5. Allow small countries to participate in the evaluation of new technologies.
6. Provide continuity of funding for coconut research.
7. Facilitate participation in an international coconut research effort by both public and private sector agencies.

The buyers of coconut oil in industrial countries would benefit from continuity of supply from producing countries. Private sector companies could be invited to participate in and contribute to an international research initiative.

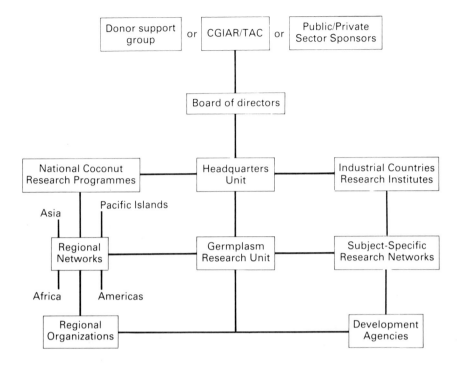

Fig. 6.1. Proposed International Coconut Research Council.

Organization

The components of the proposed International Coconut Research Council
are illustrated in Fig. 6.1. The Council would consist of the following:

Board of Directors
Headquarters Unit
 Director
 Administrative Officer
 Information Officer
 Training Officer
 Socioeconomist

Germplasm Research Unit
 Germplasm conservator – collection and conservation
 Plant breeder – hybrid production
 Plant breeder – international evaluation network
 Plant pathologist
 Research station manager

Regional Networks (four coordinators)
  Asia
  Papua New Guinea and Pacific Islands
  Africa
  The Americas

Subject-specific Networks (coordinated by above headquarters/regional staff) e.g.
  Tissue culture
  Lethal diseases
  Genetic mapping

The proposed initial complement of senior staff would be 14, plus some support staff. The Council would also require a significant contractual research budget. It would then have the responsibility to commission research of global significance with national programmes and other interested research organizations within its identified high priority research areas.

The Council would have the following functions:

1. To **identify high priority problems** of global significance, in collaboration with producing countries.
2. To **conduct research** on the conservation and utilization of coconut genetic resources. An international germplasm research unit would be established by the Council. It would be responsible for the management of a multi-site coconut germplasm collection established under international auspices, and for the interchange and evaluation of promising new hybrids and natural selections.
3. To **commission research** on the high priority problems identified. Much of this research could be undertaken by national programmes on a contractual basis. Some could also be undertaken by institutions in industrial countries, preferably in collaboration with institutions in producing countries.
4. To **sponsor selected subject-specific research networks** and foster collaborative research among scientists with common interests.
5. To **conduct and commission socioeconomic research,** particularly at the microeconomic level.
6. To **provide information and documentation** services for coconut researchers.
7. To sponsor **training programmes**.
8. To **establish four regional networks,** to facilitate the identification of priority problems and the dissemination of promising new technologies.

Board of Directors

The Board of Directors would be an international body, composed of individuals appointed in their personal capacity, who are knowledgeable about the production, problems and uses of coconut, and its research needs. The functions of the Board would be consistent with those of the CGIAR centres, which are well documented.

It would be important to include persons with knowledge of the private sector, as well as those with public sector interests. Members would come from both developing and industrial countries. One Board member should come from the Asian and Pacific Coconut Community, and another from the Bureau for the Development of Research on Tropical Perennial Oil Crops (BUROTROP), in order to ensure close collaboration with these other international agencies.

Headquarters unit

The Headquarters unit would provide the management services, training and information services, and logistical support for the germplasm research unit, the four regional networks and subject-specific research networks.

Training programme   A training programme would be associated with the priority research areas. It would include in-service training for research scientists and technicians from national programmes, and training for scientists and technicians in specialized research techniques. The training programme would cooperate with existing training institutions. It would foster study exchanges between producing countries, based on their relative strengths in different areas of coconut research. It could also include secondments to institutions in industrial countries where relevant research is being conducted. A training officer would be located at the headquarters unit to manage the programme.

Information services   Information and documentation services would be made available to coconut scientists worldwide. The proceedings of workshops and seminars sponsored by the Council would be published. The information and documentation activities would need to work cooperatively with the activities of existing agencies, especially the Asian and Pacific Coconut Community, the Coconut Research Institute in Sri Lanka, and BUROTROP. The International Development Research Centre (IDRC) commenced support in 1990 for an integrated coconut information programme, by the APCC in Jakarta. This is a significant contribution to the information needs of coconut researchers worldwide.

Socioeconomics   A socioeconomist would organize a research programme on socioeconomic issues constraining coconut production. This would include an assessment of the economic and social viability of various rehabilitation and replanting strategies.

Germplasm research unit

In the consultations which formed part of this study, it was considered essential for the effective implementation of an international research initiative on coconut that an internationally funded germplasm improvement programme be established. This would be responsible for research on collection, conservation and utilization of coconut genetic resources.

Genetic improvement is an area with high potential pay-off from research. Presently available hybrids have been shown to yield at least double the best available local material. A well-planned and internationally financed programme able to make available higher yielding material to many producing countries for evaluation under local conditions and to assist in the production of locally adapted hybrids has the potential for high returns.

The long-term nature of coconut breeding, and the need for large areas for field-based germplasm collections means that few countries have been able to mount effective breeding programmes. Even those countries which have breeding programmes suffer from lack of continuity in funding, a consequent stop–start approach to breeding, and often poor maintenance of the collections. All are breeding for national priorities. None has the responsibility to conserve the genetic resources for use throughout the world, nor to breed hybrids and make them freely available for inter-country evaluation.

The international programme on genetic improvement would require ideally the establishment of an international research unit under whose auspices coconut genetic resources could be collected and conserved (in a multisite germplasm collection) and research on techniques for germplasm conservation and breeding could be conducted. The unit would also organize intercountry evaluation of germplasm (natural selections, breeding material and hybrids).

A germplasm research unit (with five research scientists) would be responsible for the management of a coconut germplasm collection established under international auspices, with a primary site in Asia, and secondary sites elsewhere in Asia/Pacific, Africa, Latin America and the Caribbean. The unit would require sufficient land (about 1000 ha) somewhere in the Asia/Pacific region on which to plant the primary site of the international coconut genetic resources collection. Collaboration with other existing national collections would be essential, in order that parts of the international collection be replicated at sub-sites elsewhere. Existing collections, such as those in the Côte d'Ivoire and Jamaica could be key components of the multi-site collection, if those governments responsible for these collections wished to participate in the international programme. Free exchange of germplasm would be an important requirement for all participating countries.

The germplasm research unit, and the primary site of the international collection should be located ideally in an area usually free of natural disasters (especially typhoons), where there are no known lethal diseases that would preclude the exchange of material, and near the centre of origin of the crop in the Asia/Pacific region. Indonesia is one country which fulfils these criteria, and would be a suitable location at which to base an international germplasm research unit, and the primary site of the multisite international coconut genetic resources collection.

Regional networks

The function of the regional networks would be: (i) to provide guidance to the Council on the identification of problems affecting several countries, and the relative priority of these problems; (ii) to facilitate the international exchange and evaluation of germplasm; (iii) to disseminate new technologies emerging from the subject-specific research networks and commissioned research products.

It is proposed that four regional networks be established: (i) Asia, (ii) Papua New Guinea and the Pacific Islands, (iii) Africa, and (iv) the Americas.

A Regional Coordinator would act as the secretary to a Steering Committee in each geographic area. The four regional coordinators would be research workers who would also coordinate at least one of the subject-specific networks appropriate to their technical area of expertise (see below). Each could be hosted by a national coconut research institution in their geographic area of responsibility.

Subject-specific research networks

The International Coconut Research Council would also sponsor international collaborative research networks on specific subjects. These networks could be large or small, depending on their subject matter and amount of relevant research in progress. Not all member countries of the council would need to belong to all networks, since the research results would be made widely available through the regional networks.

Initial areas in which to establish international collaborative research networks could be:

- genetic resources;
- biotechnology;
- lethal diseases;
- biological control of coconut pests;
- postharvest handling and utilization;
- coconut-based agroforestry systems.

Locations of activities

It is proposed that the International Coconut Research Council establish itself in a decentralized fashion so as to facilitate participation of existing national research institutes, regional and international organizations, in the international programme, and to avoid evolving into a centralized-commodity institute.

The headquarters unit should be located in a country with substantial scientific capacity, efficient international communication facilities, and good international airline connections. In keeping with the decentralized nature of the initiative, it may be preferable for the Headquarters Unit to be located in a different country to the Germplasm Research Unit.

The Germplasm Research Unit and the main site of the international germplasm collection should be located in the centre of origin of the crop in the Asia/Pacific region, in a country with no known lethal diseases (which would preclude germplasm exchange); and outside the typhoon belt, since the primary germplasm site would be a field-based collection.

The four regional coodinators would be located in national or regional research institutes willing to provide the secretariat for the regional network. The secretariats could rotate among different institutions within a region, if more than one institute is interested in hosting the activities over time.

Options for Implementation

One of the main recommendations from the study was that the ideal outcome for an international initiative on coconut research would be to establish a new body along the lines of the 'International Coconut Research Council' (as described above). However, it was also recognized that it may not be feasible to establish a new institution (at least in the short term). In this case, it may still be possible to establish an effective international research programme on coconut which contained all the elements described above, as the programme of the 'Council', but under the auspices of an existing international agricultural research institution. Indeed, there may be some cost efficiencies and provision of common administrative services by linking the proposed international coconut research initiative with a suitable existing host institute. The goal is not to establish a new international institution *per se*, but to enable additional research to be undertaken on the global problems affecting the coconut palm.

The critical elements in comparing the options for implementation were:

- comprehensiveness of the coverage of possible research needs;
- likelihood of assembling a critical mass of scientific and financial resources;
- in-house research capacity;
- international auspices for genetic resources;
- ability to contract research on priority problems;
- likely continuity of funding;
- legal status required;
- administrative costs;
- other factors, such as the likely time for implementation.

The options considered during the consultations were ones which could be implemented within the CGIAR system, and others which could be established under international auspices, but outside the CGIAR system. The advantages and disadvantages of the selected options for the implementation of an international coconut research initiative are compared in Table 6.1

Possible host institutions include several members of the CGIAR system of international agricultural research centres. These are considered as possible hosts, either because of: their geographic location (e.g. the International Rice Research Institute (IRRI) in the Philippines); their subject matter (the International Board for Plant Genetic Resources (IBPGR), the International Council for Research on Agroforestry (ICRAF), the Centre for International Forestry Research (CIFOR)); or their mode of operation (the International Network for the Improvement of Banana and Plantain (INIBAP)).

The CGIAR system has recently expanded its scope to include research on agroforestry and forestry. ICRAF, an existing agroforestry institution located in Nairobi, has joined the CGIAR. A new Centre for International Forestry Research (CIFOR) with its headquarters unit most likely to be located in Asia is being established by the CGIAR. ACIAR is the executing agency for the establishment of CIFOR. The CGIAR also invited the International Network for the Improvement of Banana and Plantain (INIBAP), to join the CGIAR system as a new modality for supporting international agricultural research, through a 'network' approach.

The CGIAR and its Technical Advisory Committee are also considering the long-term role of the international agricultural research centres. This includes developing the concept of some of the IARC's becoming 'agroecological centres', which would undertake research on all crops of importance in a particular ecological zone. This would enable some of the existing IARCs to expand their range of commodities as they take on responsibility for a particular agroecological zone.

Other possible hosts for an international coconut research programme if established under international auspices but outside the CGIAR system are the Asian and Pacific Coconut Community (APCC) in Indonesia, or the Bureau for the Development of Research in Tropical Perennial Oil Crops (BUROTROP), with its secretariat located in France.

In the midst of the current substantive institutional and programmatic changes within the CGIAR system, it is likely to take some time for the CGIAR to resolve its views on the most suitable, long-term institutional arrangement to support a comprehensive international initiative on coconut research. In part, the scope of an international initiative will depend on the financial resources available, the interest of the management and Boards of Trustees of prospective host institutes in expanding their coverage to include

Table 6.1. Options for Implementation of an International Coconut Research Initiative.

Institutional option	Comprehensive coverage[a]	Critical mass resources[b]	In-house research capability[c]	International auspices for germplasm[d]	Contract research capacity[e]	Likely continuity of funding[f]	New legal status necessary[g]	Admin costs[h]	Other factors
1. ICRC	Yes	Yes	Yes	Yes	Yes	Yes	Yes	Medium	Possibly slow implementation
2. CGIAR/IARC	Yes	Yes	Yes	Yes	Yes	Yes	Yes	High	Establishment and governance costs high; slow implementation
3. IRRI-based unit	Yes	Yes	Yes	Possibly	Yes	Yes	Possibly	Low	IRRI mandate would require modification
4. ICRAF unit	Possibly	Possibly	Some	Yes	Yes	Yes	No	Medium	Possible linkage with ICRAF's new Asian programme
5. CIFOR unit	Yes	Possibly	Some	Yes	Yes	Yes	No	Low	Efficiencies in linkages with CIFOR, at Asian location
6. INIBAP/C network	Yes	Possibly	Possibly	Yes	Yes	Yes	Possibly	Medium	INIBAP charter may require modification
7. IBPGR genetic resources network	No	Possibly	Possibly	Yes	Yes	Yes	No	Low	Restricted to genetic resources; could include a germplasm research unit
8. APCC unit	Yes	Possibly	Possibly	Yes	Yes	Possibly	No	Medium	Location in Indonesia an advantage

Option	a	b	c	d	e	f	g	h	Comments
9. BUROTROP unit	Yes	Possibly	Possibly	No	Possibly	No		Medium	BUROTROP Charter may require modification
10. Enabling mechanism (CRC)	Yes	Possibly	No	No	Yes	Possibly	No	Low	Early implementation possible

1. International Coconut Research Council (ICRC) established independently.
2. International Agricultural Research Centre within CGIAR.
3. International Coconut Research Unit based at the International Rice Research Institute (IRRI) in the Philippines.
4. International Coconut Research Unit under the auspices of the International Council on Agroforestry (ICRAF).
5. International Coconut Research Unit under the auspices of the new Centre for International Forestry Research (CIFOR).
6. International Network for the Improvement of Banana, Plantain (and Coconut).
7. Coconut Genetic Resources Network, under the auspices of the International Board for Plant Genetic Resources.
8. Research Unit host by Asian and Pacific Coconut Community in Indonesia.
9. Research unit (or network) sponsored by Bureau for Research on Tropical Perennial Oil Crops (BUROTROP) in Paris.
10. Enabling mechanism to contract research on global problems, through a Coconut Research Consortium (CRC).

[a] Comprehensive coverage of coconut research needs, pre- and postharvest possible.
[b] Institutional mechanism would enable critical mass of expertise and resources to be directed at research needs.
[c] In-house research capacity for germplasm research unit possible.
[d] Likely availability of international auspices for multisite germplasm collections and inter-country evaluation trials in the short term.
[e] Research able to be commissioned to national research institutes in producing countries, regional organizations and advanced laboratories.
[f] Provides mechanism for continuity of funding.
[g] Need for new legal status.
[h] Administrative costs (including costs of establishment and governance).

coconut, (either because of the relevance of the subject matter, or the importance of coconut in the agroecological zone of interests), or of prospective sponsors establishing an independent initiative outside the CGIAR system.

The CGIAR agreed to TAC's recommendation that first priority be given to coconut genetic resources, and encouraged IBPGR to strengthen its work on coconut genetic resources. The progress being made by IBPGR in the establishment of a coconut genetic resources network is described in the following chapter.

TAC's recommendations on institutional arrangements saw a clear need for networks, with a strong 'enabling' component to fund research. TAC further considered that provision for this enabling component should be made through close cooperation with an existing development agency (see Chapter 5). In order to respond to TAC's recommendation as to the importance of an 'enabling component' to be associated with a research network, it is proposed that a group of interested development agencies, producing countries and other institutions establish a 'Coconut Research Consortium'. This is seen as a flexible mechanism by which to initiate activities and contract research of global interest on coconut in the short term. The next steps proposed to carry these recommendations forward are outlined in the following chapter.

Chapter seven:
Next Steps

The need is now to translate recommendations into reality

Recommendations

The critical elements of an international initiative on coconut research are: (i) to identify a set of priority problems of global significance; (ii) to establish an 'enabling mechanism' by which research on these priority problems could be undertaken on a contractual basis by scientists in national programmes, regional organizations and laboratories in industrial countries; (iii) to provide as a first priority, international auspices for a programme of coconut germplasm collecting, conservation, exchange and utilization.

The priority research areas appropriate for international support are discussed in Chapter five, and are summarized in Box 5B. First priority is given to the conservation, evaluation and use of coconut genetic resources.

It is recommended that the next steps required to establish an international initiative on coconut research are: (i) to establish an *international coconut genetic resources network*, under the auspices of IBPGR; and (ii) to establish a *coconut research consortium*.

The advantages of the two-pronged approach proposed are that: (i) the coconut genetic resources network would involve coconut researchers from all producing regions in the planning and conduct of a research programme of global significance; and (ii) the coconut research consortium would provide a focal point through which to mobilize additional funds for a carefully prepared international research programme which had broad support. The consortium would focus its support on research which is 'international in character', i.e. it is of importance to many producing countries, but beyond the scope of any one country or agency to support.

The scope of activities to be undertaken by the international coconut genetic resources network and the coconut research consortium are outlined in the following sections.

International Coconut Genetic Resources Network

Research of global interest on conservation and utilization of the genetic diversity of the coconut palm is being initiated under the auspices of an international network on coconut genetic resources. The International Board for Plant Genetic Resources (IBPGR) is presently fostering the development of such a network, under the auspices of the CGIAR. This is a first step towards establishing an international research programme on coconut germplasm improvement, as described in Chapter six.

The purposes of the network are to:

- describe the diversity
- conserve the genetic resources
- collect additional material which is at risk
- facilitate the interchange of germplasm
- evaluate different natural ecotypes and/or hybrids in different environments
- develop improved methodologies for the conservation and interchange of the genetic reources.

The approach envisaged by IBPGR for crop networks uses the principle of integration of germplasm collectors, curators, researchers, breeders and users into crop-specific networks. These crop networks are being established by IBPGR for several crops in order to ensure the wider use of germplasm collections, provide better support to crop improvement programmes, and involve a larger number of countries more closely in plant genetic resources activities.

The networks enable all aspects of germplasm collection, conservation, characterization, evaluation and exchange of plant material for a given crop to be dealt with by individuals who have a direct interest in the conservation and utilization of the species gene pool.

IBPGR and the Agency for Agricultural Research and Development (AARD) and its Central Research Institute for Industrial Crops sponsored an international workshop on coconut genetic resources in Cipanas, Indonesia in October 1991. The workshop was cosponsored by ACIAR, CIDA, CTA and IDRC. The scope of the proposed network, its membership, mode of operation, workplan and financial requirements was discussed among some 40 participants.

Participants at the International Workshop on Coconut Genetic Resources called for the implementation of an international coconut genetic resources network, and asked IBPGR to be the Executing Agency for Phase I (5 years) of this network. A comprehensive plan of action was prepared on the basis of recommendations made by three specialized Working Groups. This plan covers a contract research programme, an international database on coconut genetic resources, a strategy for the safe movement and conservation of the germplasm, and proposals for collaborative activities for better use of the genetic resources. This plan of action, unanimously endorsed by participants at the Workshop, is described in Box 7A. A budget which recommended expenditure of approximately US$1 million per year for 5 years was also prepared (IBPGR 1991, 1992).

Participants nominated a Steering Committee to develop the global 5-year programme in accordance with the recommendations of the Workshop, to plan the activities on an annual basis, and to provide the overall scientific guidance of the network. IBPGR was requested to appoint as soon as possible a full-time coordinator to be the secretary of the Steering Committee and to organize the day-to-day activities of the network.

The Indonesian meeting was critical in formulating a detailed plan of action to address some of the longstanding problems of the crop. The establishment of an international network on coconut genetic resources will enable research of global interest to be initiated in the near future.

The participants of the workshop strongly emphasized that the success-ful implementation of the coconut genetic resources network will not by itself solve the socioeconomic problem of millions of smallholders growing coconut as well as those working in the processing and marketing sectors. The need to take all socioeconomic aspects of coconut growing into account was stressed. Participants recognized, for example, that the problem of diseases and of their control could not be satisfactorily addressed by the genetic resources network alone. The Workshop called therefore for a broader initiative on coconut research, along the lines of the International Coconut Research Council outlined in Chapter six. This could include the coconut genetic resources network as one of its essential components, in the longer term (IBPGR, 1991).

Coconut Research Consortium

An important component of any international initiative would be an enabling mechanism by which research could be commissioned on an agreed set of priority problems of importance to many producing countries but which are beyond the scope of any one country to support. Thus, research within the priority areas which is 'international in character' could be ad-dressed through this mechanism. It is proposed that a 'Coconut Research Consortium' could provide this 'enabling mechanism.'

The priority problems which need to be addressed on a global basis are: (i) coconut germplasm collecting, conservation evaluation and enhancement; (ii) the major lethal diseases; (ii) the sustainability of coconut-based agroforestry systems; (iv) postharvest handling and utilization; and (v) socioeconomic issues, especially these influencing the participation of farmers in rehabilitation and replanting programmes.

The primary tasks of a Coconut Research Consortium would be: (i) to support the early establishment of the international coconut genetic resources network; and (ii) to mobilize additional financial and technical support to commission research on priority problems of global significance (possibly on a competitive basis).

The consortium could be comprised of representatives of producers and consumers, including the importers of coconut products, interested bilateral and multilateral development agencies, and other organizations active in their support of coconut research. The Coconut Research Consortium could be advised by a technical committee knowledgeable about the crop, and its research needs. One of the member agencies of the Consortium could provide the necessary secretariat services. It is proposed that the members of the group establish the necessary financial mechanisms to enable the constortium to commission research.

The advantages of the consortium would be that it: (i) builds on and supports existing research capacity in national programmes; (ii) provides a mechanism for continuity of funding; (iii) has low overhead costs; and (iv) early implementation is possible.

Conclusion

A case is presented for establishing an international research initiative on coconut. This subject has been examined since the early 1970s by several bodies interested in improving the productivity of coconut, and increasing the incomes of millions of smallholders dependent on the crop. Although the problems of the crop have been identified, and the potential returns from research appreciated, all these efforts have lapsed. The needs of the crop and the millions of people who depend on it for their livelihood have not abated in the meantime. The key problem has been lack of follow-through to establish a consortium of producing countries, importers, and development agencies, that would design and implement a high quality research program that addresses the major issues facing the crop globally, and provide the continuity of funding that is essential for a perennial tree crop such as coconut. As the next step, a consortium for coconut research needs to be established, in order to translate recommendations into reality.

BOX 7A: Recommendations of the International Workshop on Coconut Genetic Resources: Cipanis, Indonesia, October 1991

CONSERVATION AND EXCHANGE

Role of national/regional collections

1. The individual national collections are considered as the basic components of the worldwide coconut collection. Through the international database, detailed arrangements will be made for the maintenance and safe replication of the unique parts of the collections participating in the network. All countries will be invited to join the network.
2. All original material should be replicated in another collection within the conservation network.
3. The regional approach is recognized as the most appropriate way for step-by-step implementation of an international collection, which, it is foreseen, will compromise elements of national collectioins as well as regional collections in the long term. Conservation protocols will be developed for adoption by all members of the network.
4. The Workshop recommends that the Steering Committee further investigates the advantages and disadvantages of the establishment of a multi-site international collection.

Complementary conservation strategies

5. The field gene banks are recognized as a basic element in the future conservation strategy. The Workshop welcomes the fact that cryopreservation 'and pollen conservation will soon be available as complementary conservation strategies. It recommends that the network starts to conserve valuable material using cryopreservation and pollen storage as soon as these techniques are available.
6. The live preservation of dried embryos and somatic embryos appear to be emerging technologies and their use as complementary conservation strategies will have to be considered in the future. The potential of *in situ* conservation and the practicabilities of its application will require further studies by the network.
7. Curators should arrange for the indexing of their collections for the presence of virus/viroids.

Safe movement of germplasm

8. An international germplasm exchange programme will be developed as one of the activities of the network, and as a service to the international coconut community.
9. The international network should adopt the 1991 FAO/IBPGR technical guidelines for safe movement of coconut germplasm under its auspices.

Box 7A: Continued

GENETIC DIVERSITY AND UTILIZATION

Collecting activities

10. The Workshop recognizes that further knowledge is required on coconut diversity and distribution and that gaps exists in the present collections of coconut germplasm. Some geographic areas are inadequately explored. The Group agrees that a priority activity of the network is to identify these gaps by reference to existing national collections and regional programmes, to assign priorities for surveying and collecting from these areas, and to undertake the essential surveying and collecting activities. Breeders need to exploit this under-utilized diversity.

11. Some institutes emphasized the benefits derived by their programmes through the adoption of the Course Grid Collection method recommended by IBPGR. It was agreed that this method would be used as the reference basis for national programmes in their collecting strategies. When transmitting the results of its collecting missions to the international coconut database, each programme will outline any modifications to its strategy resulting from physical constraints or scientific considerations. The network will provide support for analysis of these data if the need arises.

Breeding

12. The Workshop recommends that the national programmes provide the network with comprehensive information on their on-going breeding programmes to avoid unnecessary duplication. However, it is recognized that the flow of information may be insufficient for breeders to develop standardized procedures and a minimum common strategy. It is therefore recommended that the network convene regular regional working sessions for breeders to discuss and agree on further studies related to:

- selection criteria and combining ability
- long-term breeding strategies and any other topics relevant to coconut improvement

13. The Workshop agrees on the need for comparative regional trials and for detailed assessment of the interactions between genotype and environment, and believes that these objectives can be addressed by a single programme of multilocation variety trials.

14. The Workshop endorses the conclusions reached by the BUROTROP First African Coconut Seminar at Arusha, Tanzania in February 1991 and recommends that these conclusions be extended to other coconut growing regions by establishing similar regional programmes. The objectives of multilocation trials are to test and compare the best improved varieties/hybrids produced by national programmes across a wide range of growing conditions. The group acknowledges that national breeding programmes have different aims in accordance with local constraints, markets and end-uses. Nevertheless, such comparisons will enable member countries to benefit from the breeding achievements of others and make possible a full assessment of the genotype–environment interaction.

Box 7A: Continued

15. The Workshop recognizes that there are specific physiological traits associated with yield, such as light interception, partitioning of dry matter, photosynthetic rate, nitrate reductase level and rate of transpiration. The Workshop recommends the use of these traits to investigate differences in performance of specific populations (e.g. F_1 hybrids compared with parent populations) in order to improve methods of selecting superior individual palms and populations for future breeding work.

DOCUMENTATION AND INFORMATION

International coconut database

16. The Workshop agrees on the need for a single central coconut genetic resources database to be developed for the initial stage of the network. The offer by CIRAD Montpellier to act as a host for this database was gratefully accepted. The Workshop recommends that once the database becomes will established, the possibilities of decentralizing responsibilities for further maintenance and extension of the global database to the regional level will be examined by the Steering Committee.

17. All data that will be incorporated into the database and the genetic material to which it refers is to be freely available to network participants.

18. It is recommended that in addition to the support that the designated host of the database will provide for the compilation, distribution and maintenance of the database, additional funding should be obtained for operation of the database.

19. Owing to the lack of information on the coconut databases in participating institutes, the Workshop recommends that representatives from institutes that have established databases meet at the location of the host with copies of their data as soon as possible to determine the descriptors that will form the basis of the international coconut database. This will also allow the database to be established in their presence and difficulties to be resolved more quickly. The results of this meeting will be reported to the Steering Committee as well as recommendations for further action.

20. Some coconut collections will need assistance to develop documentation systems. The identification, assessment and development of these programmes should be undertaken on a regional basis. This will include hardware acquisition/development, software use and development, and training in documentation. These activities are planned to take place after the initial development of the international coconut database.

21. It is recommended that the development of software for use by coconut collections be examined after the finalization of descriptors and establishment of the central coconut database. This will be freely available to all participants and will help to provide standardization of descriptors and aid in the exchange of data.

Box 7A: Continued

Descriptor list

22. As the systematic description of coconut has not been undertaken internationally it is recommended that a revised descriptor list be published as soon as possible after consideration by the Steering Committee. This descriptor list should be used for a number of years and then revised.

23. It is recommended that the Steering Committee provide input into the draft descriptor list for the standardization of techniques, colour chart usage and codes, periods of observation and growth stages and that these are passed on to IBPGR for incorporation into the descriptor list as soon as possible.

Directories

24. The Workshop was informed of BUROTROP's efforts to produce a comprehensive Directory of coconut research activities. Participants also leaned about the intentions of the organizers of our international seminar in India in November 1991 to publish a comprehensive list of workers in the field. It is recommended that the network, through IBPGR, liaise with both the above-mentioned organizations to assess the need for any complementary action.

Further circulation of information

25. In addition to the regular flow of information between the central database and the members of the network, and from the Steering Committee to members of the network, the Workshop agrees on the need for a wider forum to circulate information regularly on the activities of the coconut genetic resources network. The systematic provision of space for this purpose within the BUROTROP Newsletter was regarded as the optimum solution.

RESEARCH

Conservation and safe movement

26. The Workshop identified needs for further research in the following conservation techniques: cryopreservation, pollen storage, dried embryos and somatic embryos. It recommends that high priority be given to research on new *in vitro* techniques that will ensure complementary conservation methods.

27. The development of safe indexing techniques for cadang-cadang-like viroids and for mycoplasma-like organisms is of high priority for the safe movement of germplasm. Research in this area is an essential part of the germplasm exchange programme.

28. Similarly the identification of threats to genetic resources from diseases of unknown aetiology should be one of the priority activities of the network.

29. The Workshop emphasized that it is vital for the network to be able to respond quickly to emerging research issues (e.g. concerning conservation methods and safe movement of the germplasm). Provision has been made in the budget for this purpose.

Box 7A: Continued

Genetic diversity and utilization

30. Considering the urgent need to analyse the genetic diversity in coconut for proper collecting, conservation and better utilization, and considering the relative limitations of enzymatic analysis in this field, the Workshop gives priority to the use of modern molecular techniques for greater efficiency. The network should contact specialized laboratories to make proposals for comprehensive projects dealing with this area.

31. Steps should be taken to improve the recording of all existing information on past and ongoing observations related to adaptation traits, taking into consideration environmental conditions, historical background, etc. A small task force of two to three scientists including a statistician and a physiologist will analyse this preliminary information and will recommend the collation of physical data if it appears that these can be exploited to good purposes.

32. Recognizing the value of biochemical and molecular information as indicators of genetic distance between populations, participants recommend the use of these techniques to aid the development of new F_1 hybrids.

33. Recognizing the potential value of RFLP markers for important traits such as drought tolerance, pest and disease resistance, it is recommended that assistance be sought to identify such markers so that these traits might be more rapidly incorporated into superior varieties.

TRAINING AND MANPOWER DEVELOPMENT

34. The network needs to ensure that all member countries have appropriately trained staff at all levels to fulfil its plan of action.

35. Extensive needs for training were identified in the areas of disease indexing, embryo culture techniques and the maintenance of collections. Training in this area is an integral part of the coconut germplasm exchange programme.

36. Training on all aspects of documentation, especially data acquisition and data management, should be undertaken.

37. Concerning utilization, training needs were identified at three levels: technicians, junior researchers and senior researchers. It was agreed that the first priority should be technicians. Four centres (India, Philippines, Vanuatu and Côte d'Ivoire) are now in a position to offer this type of training. However, there are some discrepancies in the techniques used and it is advisable to standardize such training at the international level. It is recommended that these four centres, with the collaboration of other scientists as required, produce a guidebook for basic germplasm operations.

38. Participants agree that training mechanisms should remain flexible. In certain conditions, training can be best achieved by sending a senior scientist to the country where training is needed. Alternatively trainees may go to the centre of expertise to work on an individual or group basis. In other cases, the production of training materials may be a prerequisite to further action.

Box 7A: Continued

39. Finally, the Workshop recommends that the Steering Committee set up a special fund in the future to provide a few study grants for PhDs linked to the network's research activities.

Source: IBPGR (1991, 1992)

References

ACIAR (1988) *Potential Australian Market for Coconut and Coconut Products.* ACIAR Working Paper No. 22, ACIAR, Canberra.

APPC (1986) *Coconut Statistical Yearbook 1985.* Asian and Pacific Coconut Community, Jakarta, Indonesia.

APCC (1989) *Report of COCOTECH Meeting, Bangkok, May 1989.* Asian and Pacific Coconut Community, Jakarta, Indonesia.

APCC (1990) *Coconut Statistical Yearbook 1989.* Asian and Pacific Coconut Community, Jakarta, Indonesia, 316 pp.

Banzon, J.A. and Velasco, J.R. (1982) *Coconut Production and Utilization.* Philippine Coconut Research and Development Foundation, Manila, 351 pp.

BUROTROP (1991a) Working towards a better future for the African coconut farmer. *Proceedings of the First African Coconut Seminar, 4–5 February 1991,* Arusha/Dar es Salaam, Tanzania. BUROTROP, Paris.

BUROTROP (1991b) Brochure. Bureau for the Development of Research on Tropical Perennial Oil Crops, Paris.

CGIAR (1986) *CGIAR Priorities and Future Strategies.* Consultative Group on International Agricultural Research Technical Advisory Committee, TAC Secretariat, FAO, Rome, 246 pp.

CGIAR (1990) *A Possible Expansion of the CGIAR.* Consultative Group on International Agricultural Research Technical Advisory Committee, TAC Secretariat, FAO, Rome.

CGIAR (1992) *CGIAR Priorities and Strategies.* Consultative Group on International Agricultural Research Technical Advisory Committee. TAC Secretariat, FAO, Rome.

Calvez, C., Julia J.F. and de Nuce de Lamothe, M. (1985) Improvement of coconut in Vanuatu and its importance for the Pacific region. *Role of the Saraoutou station, Oleagineux* 40, 477–490

Child, R. (1974) *Coconuts.* Longman, London.

Corley, R.H.V. (1983) Potential productivity of tropical perennial crops. *Experimental Agriculture* 19, 217–237.

Davis, J.S., Oram, P.A. and Ryan, J.G. (1987) Assessment of agricultural research priorities: An international perspective. *ACIAR Monograph No. 4*, 85 pp.

Enig, M. (1990) Let's talk fats and cholesterol. *US Council for Coconut Research and Information*, Washington, D.C.

Etherington, D. (1988) *A Policy Perspective on Coconut Processing for the South Pacific Countries.* CORD 4 (2), Asian and Pacific Coconut Community, Jakarta, 40 pp.

FAO (1990) *FAO Production Yearbook.* United Nations Food and Agriculture Organization, Rome.

FAO (1991) *Research and Development Program on Coconuts and their Products.* Commodities Division, FAO, Rome. CCP: OF 91/S.

FAO/IBPGR (1992) Guidelines for the safe movement of coconut germplasm. *FAO Plant Protection Service*, and *International Board for Plant Genetic Resources.* Rome (in press).

FAO/UNDP (1969–1981) Yearly progress reports on coconut breeding. *FAO Technical Working Party on Coconut Production, Protection and Processing.* FAO, Rome.

Foale, M.D. (1987) Coconut germplasm in the South Pacific Islands. *ACIAR Technical Reports Series No. 4*, Australian Centre for International Agricultural Research, Canberra, 23 pp.

Foale, M.A. (1988) An annotated bibliography on coconut research relevant to the Pacific Islands. *ACIAR Working Paper No. 10.* ACIAR, Canberra.

Foale, M.A. (1990) An annotated bibliography on the coconut palm. *ACIAR Working Paper No. 29.* ACIAR, Canberra.

Hagen, D.L. (1991) *Energy Systems for Smallscale Coconut Processing.* Report of COCOTECH XXVII, Suva, Fiji, July 22–26, 1991. APCC, Jakarta.

Hanold, D. and Randles, J.W. (1991a) Coconut cadang-cadang and its viroid agent. *Plant Disease* 75(4), 330–335.

Hanold, D. and Randles, J.W. (1991b) Detection of coconut cadang-cadang viroid-like sequences in oilpalm and coconut palm and other monocotyledons in the south west Pacific. *Annals of Applied Biology*, 118.

Harries, H.C. (1978) The evolution, dissemination and classification of *Cocus nucifera* L. *Botanical Review* 44, 265–320.

Harries, H.C. (1991) *Two Hundred and Fifty Years of Coconut Research.* Cocomunity XXI No. 19, II.

Harries, H.C. (1992a) Coconut *Cocos nucifera* L. *In*: Smart, J. (ed.) *Simmonds Evolution of Crop Plants* (2nd edn) (in press).

Harries, H.C. (1992b) Coconut. *In*: Macrae, R. (ed.) *Encyclopaedia of Food Science and Nutrition.* Academic Press, London (in press).

Harvard Medical School (1987) Testimony before the US Senate Committee on Labor and Human Resources on S. 1109, a bill to amend the Federal Food, Drug and Cosmetic Act to require new labels on foods containing coconut, palm, and palm kernel oil. *Coconuts Today*, December 30, 1987, pp. 22–23

Howard, F.W. (1991) Ecology and control of hemipherous pests of cultivated palms. *American Entomologist*, Winter 1991, 217–225.

IBPGR (1976) *Report of Working Group on Coconut Genetic Resources*. 27–29 April 1976. International Board for Plant Genetic Resources, Rome.

IBPGR (1981) *Directory of Germplasm Collections*. 5.1 Industrial Crops. International Board for Plant Genetic Resources, Rome.

IBPGR (1991) *Report of an International Workshop on Coconut Genetic Resources*, Cipanas, Indonesia. October 8–11, 1991, IBPGR, Rome, 18 pp.

IBPGR (1992) *Proceedings of an International Workshop on Coconut Genetic Resources*, Cipanas, Indonesia. October 8-11, 1991, IBPGR, Rome (in press).

IBPGR/SEAP (1987) *Report of Working Group on Palms and Genetic Improvement*. International Board for Plant Genetic Resources, Rome.

IRHO (1986) *Creation of a Research Organization for Tropical Perennial Oil Crops*. Institut de Recherches pour les Huiles et Oleagineaux, Paris, 3 volumes.

Johnson, D.V. and Nair, P.K.R. (1989) Perennial crop-based agroforestry systems in northern Brazil. *In*: Nair, P.K.R. (ed.) *Agroforestry Systems in the Tropics*. Kluwer Academic Publishers, Dordrecht, The Netherlands, pp. 475–487.

Jones, J.M. (1989) Tropical oils: Truth or consequences. *Cereal Foods World* 34 (10) 866–868 870–871.

Jones, L.H. (1990) Oil palm and coconut. *In*: Persley, G.J. (ed.) *Agricultural Biotechnology: Opportunities for International Development*. CAB International, Wallingford, UK, pp. 213–224.

Liyanage, M.deS., Tejwani, K.G. and Nair, P.K.R. (1989) Intercropping under coconuts in Sri Lanka. *In*: Nair P.K.R. (ed.) *Agroforestry Systems in the Tropics* Kluwer Academic Publishers, Dordrecht, The Netherlands, pp. 165–179.

Mielke, S. (1988) Oilseeds, oils, fats and meals: Tendencies since 1958 and perspectives up to the year 2000. *Oilworld*, 11 March 1988.

Nair, P.K.R. (1983) Agroforestry with coconuts and other tropical plantation crops. *In*: Huxley, P.A. (ed.) *Plant Research and Agroforestry*, ICRAF, Nairobi, pp. 79–102.

Nair, P.K.R. (ed.) (1989) *Agroforestry Systems in the Tropics*. Forestry Sciences, Vol. 31. Kluwer Academic Publishers, Dordrecht, The Netherlands, in cooperation with ICRAF, Nairobi, Kenya.

Ohler, J.G. (1984) Coconut, tree of life. *FAO Plant Production and Protection Paper 57*. FAO, Rome, 446 pp.

Oilworld, 1992. Demand situation and prospects. *Oilworld*, 36(14), 108.

PCA (1986) *Spectrum of Coconut Products II*. Philippines Coconut Authority, Quezon City, Philippines.

Persley, G.J. (1988) Coconut Research: An International Initiative. Paper presented to the CGIAR Technical Advisory Committee, June 1988. TAC Secretariat, FAO, Rome, 64 pp.

Persley, G.J. (1989) Coconut: International Research Priorities. Paper presented to the CGIAR Technical Advisory Committee, October 1989. TAC Secretariat, FAO, Rome, 73 pp.

Persley, G.J. (1990a) The Coconut Palm: Prosperity or Poverty. Paper presented to the CGIAR Technical Advisory Committee, June 1990. TAC Secretariat, FAO, Rome, 81 pp.

Persley, G.J. (1990b) Coconut Research Opportunities. Paper presented to the CGIAR Technical Advisory Committee, June 1990. TAC Secretariat, FAO, Rome, 29 pp.

Persley, G.J. (1990c) *Beyond Mendel's Garden: Biotechnology in the Service of World Agriculture.* CAB International, Wallingford, UK, 155 pp.

Persley, G.J., Foale, M.A. and Wright, B. (1990) *Coconut Cuisine.* Inkata Press, Melbourne, 57 pp.

Pieris, W.V.D. (1968) *Introduction and Exchange of Coconut Germplasm 1959 to 1966.* Report of 3rd FAO Technical Working Party on Coconut Production, Protection and Processing, Jakarta, Indonesia. FAO, Rome.

Plucknett, D.L. (1979) *Managing Pastures and Cattle under Coconuts.* Westview Press, Boulder, Colorado, USA, 364 pp.

Punchihewa, P.G. (1991) Objectives and functions of the Asian and Pacific Coconut Community. *In: Proceedings of the First African Coconut Seminar,* 4-8 February 1991, Tanzania. BUROTROP, Paris, pp. 14–17.

Purseglove, J.W. (1975) *Tropical Crops: Monocotyledons.* Longmans, London, pp. 440–479.

Randles, J.W., Hanold, D. and Julia, J.F. (1987) Small, circular single-stranded DNA associated with foliar decay disease of coconut palm in Vanuatu. *Journal of General Virology* 68, 273–280.

Robertson, J.S., Prendergast, A.G. and Sly, J.M. (1968) Diseases and disorders of the oil palm (*Elaeis guineensis*) in West Africa. *Journal of the Nigerian Institute for Oil Palm Research* 4, 381–409.

Rohde, W., Randles, J.W., Langridge, P. and Hanold, D. (1990) Nucleotide sequence of a circular single-stranded DNA associated with coconut foliar decay virus. *Virology* 176, 648–651.

Schirmer, A. (ed.) (1984) *The Role of Agroforestry in the Pacific.* DSE (German Foundation for International Development), Berlin, Germany.

Smith, M.A. and Whiteman, P.C. (1983) Evaluation of tropical grasses in increasing shades under coconut canopies. *Experimental Agriculture* 19, 153–161.

UNDP/FAO (1988) *Report of Working Group on Genetic Improvement,* Chumphon, Thailand. FAO, Rome.

Vergara, N.T. and Nair, P.K.R. (1989) Agroforestry in the South Pacific region; an overview. *In:* Nair, P.K.R. (ed.) *Agroforestry Systems in the Tropics.* Kluwer Academic Publishers, Dordrecht, The Netherlands, pp. 291–306.

Waterhouse, D.F. and Norris, (1987) *Biological Control of Insect Pests in the South Pacific.* Inkata Press, Melbourne.

Way, M. (1992) Role of ants in pest management. *Annual Review of Entomology* 37, 479–503.

Whitehead, R.A. (1966) *Sample Survey and Collection of Coconut Germplasm in the Pacific Islands* (30 May–5th September 1964). Ministry of Overseas Development, HMSO, London.

Wilson, G. and Ludlow, M. (1991) The environment and potential growth of herbage under plantations. *In:* Shelton, H.M. and Sturl, W.W. (eds). *Proceedings of the International Workshop on 'Forages for Plantation Crops'.* Udayana University, Indonesia, June 1990. ACIAR, Canberra.

World Bank (1988) Price prospects for primary commodities, Report No. 814/88, Vol. II: *Food Products and Fertilizers, and Agricultural Raw Materials.* World Bank, Washington, D.C.

World Bank (1991) Price prospects for primary products. Vol. II: *Food Products and Fertilizers, and Agricultural Raw Materials.* World Bank, Washington, DC.

World Bank (1991) Coconut production: Present status and priorities for research. *World Bank Technical Paper Number 136*, Washington DC, 150 pp.

Wright, B. and Persley, G.J. (1988) Coconut: the tree of life that is slowly dying. *Partners in Research for Development*, No. 1. ACIAR, Canberra, pp. 10–13.

Index